Advanced Mechatronics

Monitoring and Control of Spatially Distributed Systems

Advanced Mechatronics

Monitoring and Control of Spatially Distributed Systems

Dan Necsulescu
University of Ottawa, Canada

NEW JERSEY • LONDON • SINGAPORE • BEIJING • SHANGHAI • HONG KONG • TAIPEI • CHENNAI

Published by

World Scientific Publishing Co. Pte. Ltd.
5 Toh Tuck Link, Singapore 596224
USA office: 27 Warren Street, Suite 401-402, Hackensack, NJ 07601
UK office: 57 Shelton Street, Covent Garden, London WC2H 9HE

British Library Cataloguing-in-Publication Data
A catalogue record for this book is available from the British Library.

ADVANCED MECHATRONICS
Monitoring and Control of Spatially Distributed Systems

ISBN-13 978-981-277-181-0
ISBN-10 981-277-181-6

Printed in Singapore.

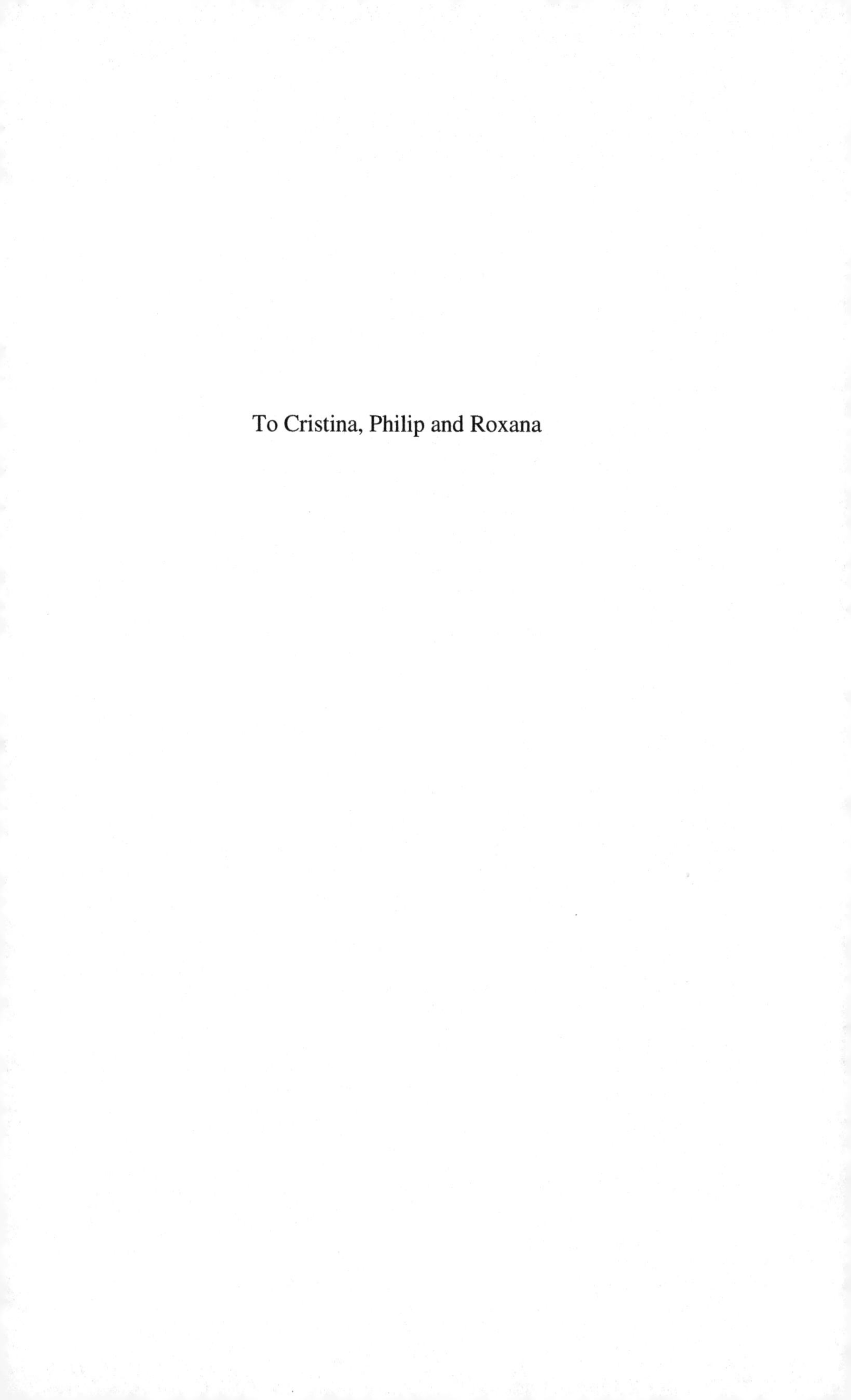

To Cristina, Philip and Roxana

Preface

Mechatronics is an engineering field that refers to mixed systems tight integration. Currently, this integration can be viewed as based on digital computer monitoring and control, but it cannot be denied that integration can be based on any other signal processing system and any form of raw power that can be modulated and transferred to the mixed system in accordance to the output of this digital signal processor.

Distributed parameters systems, in the form of solids, liquids, gases, are seen as fields in which the dynamics can be represented by traveling waves. These fields can be mechanical vibration fields of substance, acoustic, electromagnetic etc. The assumption of continuity is often limited, when moving down from macroscopic level, by the molecular, atomic discontinuous structure, which can be represented in duality with the fields at that level. Moving up from immediate macroscopic level towards infinite celestial level, again the continuum of the quasi-vacuum space is filled with solid planets etc. Consequently, while at terrestrial macroscopic level, continuous fields can be assumed of infinite dimensions, there are perceived limits as we move up and down from this level.

Distributed Parameters Systems are modeled mathematically by partial differential equations and/or multiple integrals that can be recast also in a system of partial differential equations. The solutions of these partial differential equations show that the dynamics of distributed parameters systems can be simulated as composed of infinite dimensional combinations of harmonic components (something that might remind of Pythagoras' view of the planets motions) where higher frequency components might become less and less significant unless excited and brought to resonance.

Mechatronics refers to monitoring, control and integration not only of lumped parameters systems, but also of distributed parameters systems. In fact, the latter representation of the world under engineering focus is more realistically modeled by distributed parameters systems; handling such models is, however, much more difficult that the lumped parameters systems. Monitoring and control of distributed parameters systems is limited by ill-posed problems, the inverse problems of estimating system states and parameters from sensors signals and controlling and infinite dimensional system with modulated power output from actuators. Sensors and actuators are available in most cases as point devices and, even if they are distributed, they cannot be found in the infinite dimensional form. Sensors and actuators are bandwidth limited and cannot access higher frequency components of distributed parameters systems dynamics. As a result, only lower frequency dynamics can be controlled and maybe somewhat higher but still low frequency components can be monitored; higher frequency dynamics remains uncontrolled and unobserved. Pascal made a valid comment with regard to human condition in an infinite world: "...qu'est que l'homme dans la nature? Un néant à l'égard de l'infini, un tout à l'égard du néant, un milieu entre rien et tout." (B. Pascal, Pensées, no. 72). Using science and engineering, we reach easily documented limits in monitoring and controlling such systems and only religion, art and philosophy can offer further views outside these limits. Indeed, direct view, i.e. intuitive access to that level requires to become detached from contact and affection from the immediate and finite environment and to bring ourselves to the vision of infinite spaces.

I acknowledge the results documented in the book of joint published research with my colleagues professors Dr. R. Baican F. Bakhtiari-Nejad, J. Sasiadek and W. Weiss and with my former graduate students: R. F. De Abreu, G. M. Ceru, G. Ganapathy, Kuoc-Vai Iong, Y. Jiang and W. Zhang.

Dan Necsulescu, Ottawa, Canada

Contents

Chapter 1

Introduction

1.1 Advanced Mechatronics Systems

1.1.1 *Monitoring and Control of Distributed Parameters Systems*

Most engineered systems are composed of mixed mechanical-electrical-electronic-thermal subsystems and have fewer sensors (under-sensing) than states needed for monitoring and control and, moreover, have fewer actuators than degrees of freedom (under-actuated). Some of these systems can be modeled in a first approximation as lumped parameters systems but, in general, require more complex approaches for proper design and operation.

The focus in this Advanced Mechatronics text is on the computer based -integration, -monitoring and -control of mixed systems that can be described as distributed parameters systems. The illustrations for distributed parameters systems will be acoustic fields, thermo-dynamic fields, magnetic fields, vibrations in flexible structures, *etc*. The following topics will be presented:

- overview of advanced mechatronic systems: signals versus power transmission, local sensing and actuation in continuous systems, centralized versus local control
- modeling and control issues for mixed systems: effort-flow modeling, modeling and simulation of distributed parameters systems, open and closed loop control
- numerical solutions for inverse problems using regularization and singular value decomposition methods
- dynamic calibration of sensors

- transient response of under-actuated and under-sensed systems
- active vibration control in flexible structures
- acoustic fields monitoring and control
- thermo-dynamic fields in thermal process control
- magnetic fields in magnetic levitation.

Figure 1.1 shows the schematic diagram of a distributed parameters mechatronic system. In Fig. 1.1 system variables are measured by transducers, signal conditioned and converted from analog to digital form and transmitted to a computer. The computer performs real time monitoring and control as well as signal analysis and has two types of outputs, one for actuator commands and the other for system monitoring display.

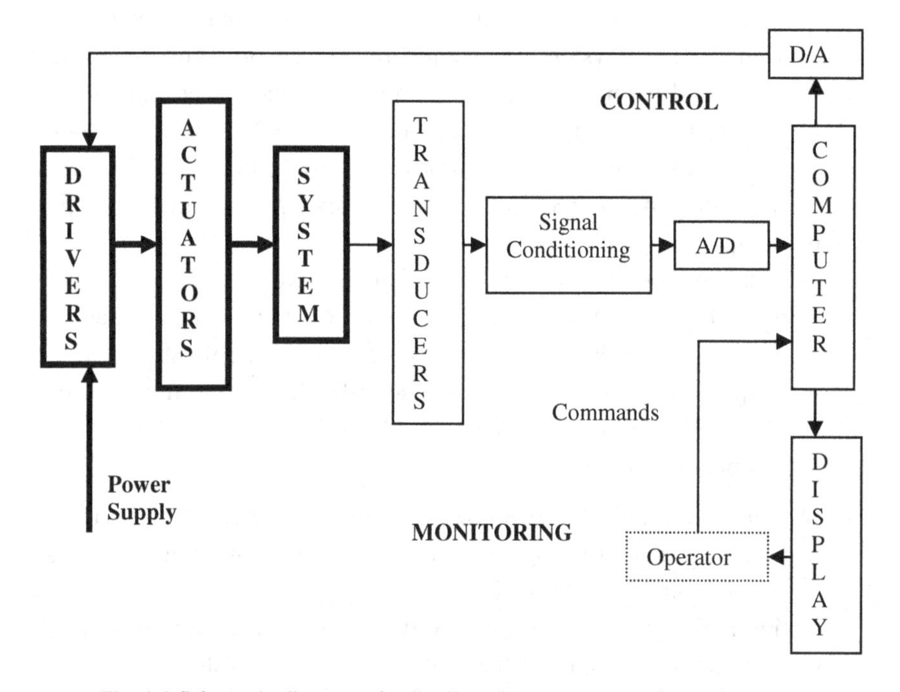

Fig. 1.1 Schematic diagram of a distributed parameters mechatronic system

The commands are either operator commands, or computed commands that are shown applied in a closed loop control configuration.

Computer output for control, after conversion from digital to analog form, sends commands to actuators.

Control commands are signals sent to drivers that modulate the power from an external power supply for the actuators.

An advanced mechatronics approach has to take into account that physical systems are inherently distributed parameters systems and that only some of these systems can be represented by a lumped parameters model. Lumped parameters mechatronic systems were already investigated extensively in several mechatronics books [1-9]. Figure 1.1 refers to a distributed parameters mixed system that can represented by partial differential equations [25, 44, 110]. Numerous distributed parameters systems are mixed systems. Examples analyzed in this text are: acoustic, thermal, fluid, magnetic systems and flexible structures.

1.2 Signals versus Power Transmission. Lumped Parameters Modeling of Mechatronic Systems

Integration of systems is achieved transmission of signals and power between subsystems.

Distributed parameters systems modeling require modeling of propagation delays, boundaries effects, 3D interactions etc, which are not present in a lumped parameters model or in its block diagram counterpart. Lumped parameters systems, described by Linear Time Invariant (LTI) Ordinary Differential Equations (ODE), are reviewed in this section, in order to identify specific needs for modeling distributed parameters systems.

Block diagrams contain variables associated to the unidirectional links between blocks. These variables can be seen as signals containing the information transmitted from the output of one block to the input of another block. In control engineering signal flow graphs are sometimes used as an equivalent alternative form to block diagrams.

What is important in communication systems is only the information contained in the signals, not the power transmitted by the carrier of this information. In this case, blocks represent transformations applied to the signal transmitted, for example delays, attenuation or filtering. On a communication link, signals can be transmitted bi directionally. Block diagram models represent only unidirectional transmission, from the designated output of one block to the designated input of another block and consequently contain only a direct model, from "cause" to "effect". The model associated with a block corresponds only to the transfer from the input to the output. This might be acceptable for signal transmission, but for power transmission, which is normally bidirectional, effort-flow models are required.

Inverse model, from desired output to the required input, is obtained by matrix inversion for square LTI systems. Inverse model for non-square LTI systems require pseudo-inverse. For non-linear systems, no closed form solution might be available for model inversion.

In other engineering systems, the power transmitted by the carrier becomes important, and the equations describing their dynamics are written for variables like force and velocity in mechanical systems and voltage and current in electrical circuits. Equations using these variables can also be used for block diagram modeling. Again, while power often flows bi-directionally on a transmission line, a block diagram model can represent only a single direction of the transmission. In fact, state space models, transfer function and block diagram representations require the assignment of the direction of the signal from one component of the model to another.

Example 1.1 Consider first a simple mechanical system example, shown in Fig. 1.2, composed of a mass m, a spring k and a damper b and subject to a force input F. The velocity v is assumed the output.

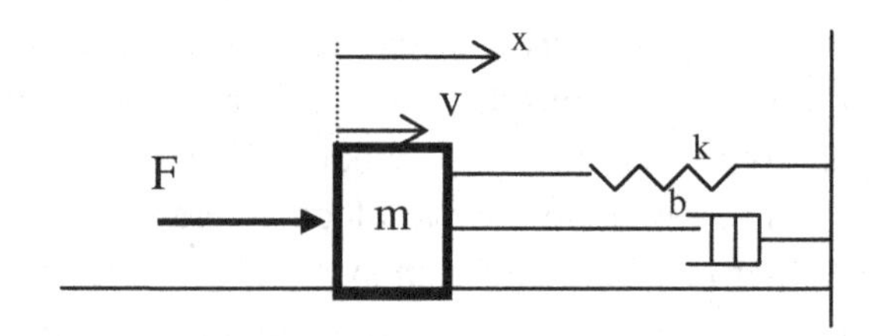

Fig. 1.2 A mass-spring-damper example

Newton's second law gives

$$F = m\frac{d^2}{dt^2}x(t) + b\frac{d}{dt}x(t) + k \cdot x(t)$$

The above differential equation can be written using v as variable

$$F = m \cdot \frac{d}{dt}v(t) + b \cdot v(t) + k \cdot \int_0^t v(\tau)d\tau$$

for

$$v(t) = \frac{d}{dt}x(t)$$

Laplace transform for zero initial conditions gives

$$v(s) = \frac{1}{ms + b + k/s}F(s)$$

Due to the input and output assignments, the same system is modeled differently when the variables F and v change designation. In this case, a simple inversion of the transfer function gives the inverse model

$$F(s) = (m \cdot s - b - k/s) \cdot v(s)$$

In general, however, model inversion does not have a closed form solution, typically for distributed parameters systems. This restricts modularity and interchangeability to modules with identical input and output assignment.

1.2.1 *Effort Flow Variables and Two Port Models*

Two port models were introduced for representing components of electric networks using two terminals for each port. Alternative names for two port components of a network are: four terminal network or quadripole. The two pole port models have associated a current I variable and a voltage V variable that permit the calculation of the power P = VI, transferred through the port [8,9].

Example 1.2 For an inductance-resistance L-R circuit supplied by an ideal voltage E source (i.e. with zero internal impedance), the circuit is shown in Fig. 1.3. Obtain the tree cuts diagram.

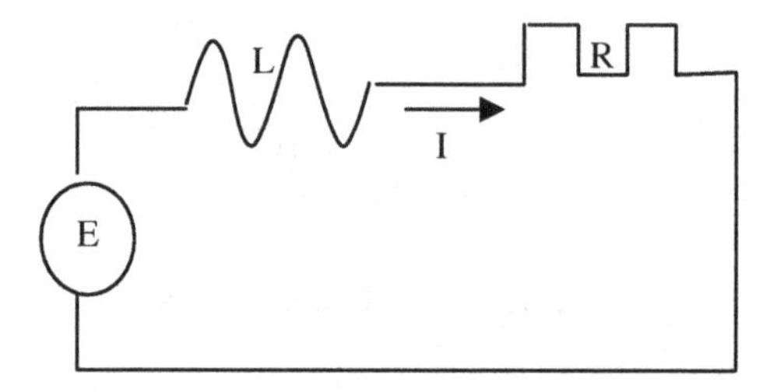

Fig. 1.3 A resistance-inductance R-L circuit

Resistance R and capacitance L components can be represented as separate elements as a result of three cuts (Fig. 1.4).

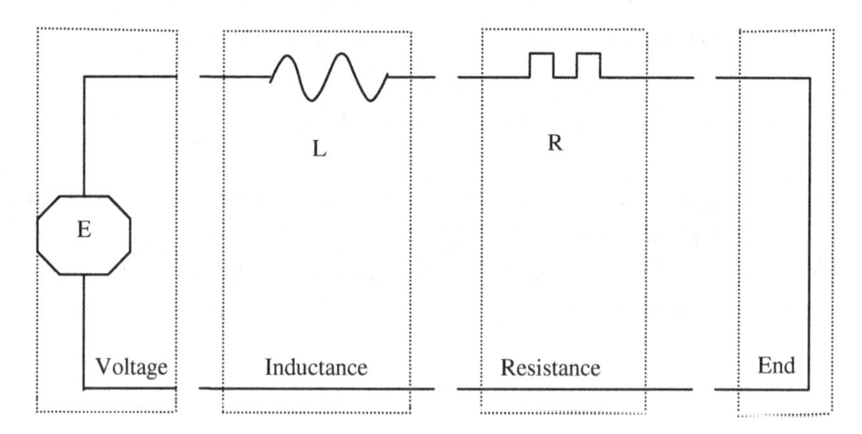

Fig. 1.4 Three cuts in R-L circuit

Example 1.3 A resistance - inductance-capacitance (R – L – C) series circuit subject to voltage V is shown in Fig. 1.5. Obtain Z(s) = V(s) / i(s)

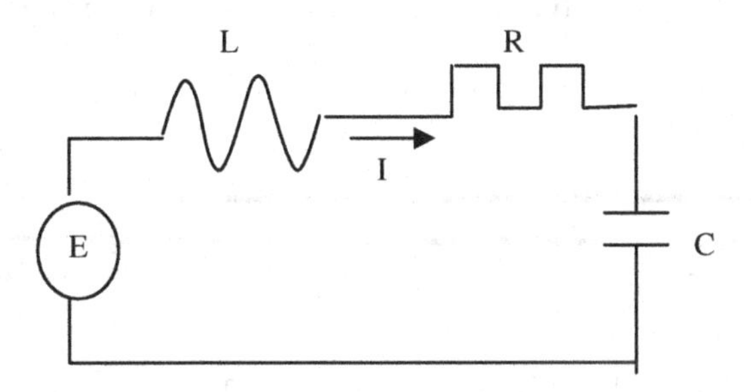

Fig. 1.5 R-L-C circuit

Solution The following voltage drop equation can be written

$$V(t) = Ri(t) + L\frac{d}{dt}i(t) + C\int_0^t i(\tau(\tau)$$

Laplace transform of the above equation for zero initial conditions gives

$$V(s) = (R + Ls + (C/s)) \cdot i(s)$$

The impedance Z(s) of the resistance – inductance – capacitance series circuit is given by

$$Z(s) = V(s) / i(s)$$

or

$$Z(s) = R + L \cdot s + C/s$$

In the case of solid body mechanics, free body diagrams represent components of a multi body system obtained by cutting each "body" from the system and representing boundary effects by local force f and velocity v whose product gives the power P = f · v.

Example 1.4 Assume that a flexible horizontal rod is linked to an undefined arbitrary mechanical systems by spherical joints. The rod, cut from these systems, give the free body diagram shown in Fig. 1.6. Obtain the model.

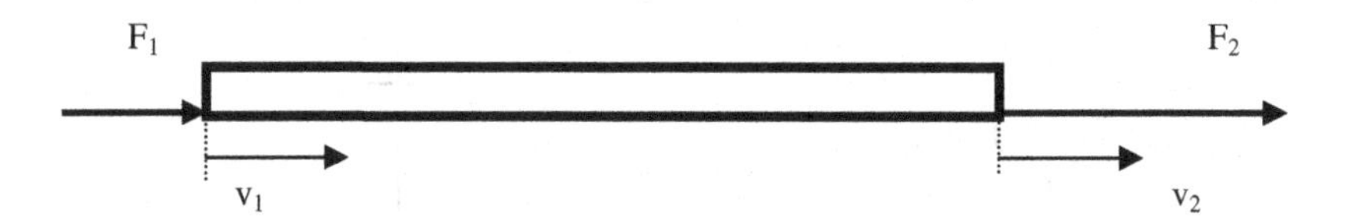

Fig. 1.6 Free body diagram of a rod

In Fig. 1.6 for each cut the internal force F and the absolute velocity v are identified. Assuming the flexible rod represented by a lumped parameters model, shown in Fig. 1.7, the following equations can be written

$$F_1(t) = (x_1(t) - x_2(t)) \cdot k + (v_1(t) - v_2(t)) \cdot b$$

$$F_2(t) = -[(x_1(t) - x_2(t)) \cdot k + (v_1(t) - v_2(t)) \cdot b] = -F_1(t)$$

Even if there is a spring and a damper between the two forces, the equality $F_2(t) = -F_1(t)$ reflects the fact that, in this model, the time-varying force change $F_1(t)$ applied to end 1 appears transmitted instantaneously at end 2, given that lumped parameters models do not account for propagation delay.

For

$$v_1(t) = \frac{d}{dt} x_1(t)$$

$$v_2(t) = \frac{d}{dt} x_2(t)$$

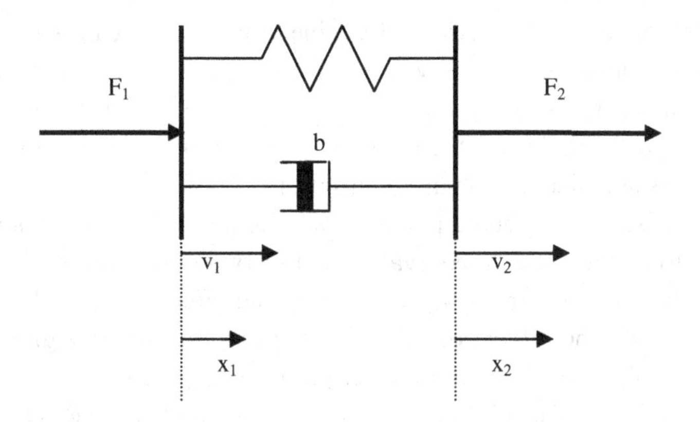

Fig. 1.7 Free body diagram for the rod

the Laplace transform gives

$$x_1(s) = v_1(s)/s$$
$$x_2(s) = v_2(t)/s$$

The two force equations give

$$v_1(s) = -[1/(k / s + b)] \cdot F_2(s) + v_2(s)$$

$$F_1(s) = -F_2(s)$$

These last two equations give the cut variables at end 1, F_1 and v_1 function only of cut variables at end 2, F_2 and v_2, and parameters b and k, i.e. independent of the dynamics of the systems to which cut 1 and 2 were applied. Lumped parameters mechanical systems can be sectioned by cuts into subsystems interfaced only by force and velocities defined with respect to the cuts. For a flexible torsional shaft, with cut parameters torque T and angular velocity ω, the model is structurally similar:

$$\omega_1(s) = -[1/(k / s + b)] \cdot T_2(s) + \omega_2(s)$$

$$T_1(s) = -T_2(s)$$

A generalization to a variety of engineering systems can based on the two port components that have associated a flow or through variable “f” and an across or effort variable “e” giving the power as the product (flow) · (effort) [8]. This is the power passing through the junction of two components associated to a particular port [9].

In the case of distributed parameters systems, the interactions are too complex to be reducible to equivalent simple two-port models.

The direction of the power flow in the junction is bidirectional as opposed to the block diagram description in which signals have unidirectional flow. The same description, using effort-flow two pole ports, is suitable for mixed systems. The theoretical background of this description can be found in Hamiltonian dynamics for obtaining power transfer equations [8]. While effort-flow cuts permit to define power transfer between mixed subsystems, Hamiltonian and Lagrange dynamics permit simultaneous modeling of mixed systems, for example of electromechanical systems [9].

1.2.2 *Newton-Euler and Kirchhoff Equations for a Mixed Electro-Mechanical System*

Effort-flow representation of mixed systems permits easy application of Newton-Euler equations of motion and Kirchhoff equations for electric circuits. Power transfer conservation law at the conversions of electrical and mechanical energies permits to integrate the two models in an joint electro-mechanical model. The simplified diagram of Permanent Magnet-Direct Current (PM-DC) motor is shown in Fig. 1.8. The stator consists of a pair of magnetic poles N-S. The rotor consists of coils of conducting wires connected through the segments of a collector to a DC power supply.

Figure 1.8 shows the cut from a mechanical load (with cut variables torque T and angular velocity $\omega = d\theta/dt$) as well as the cut from a DC power supply (with cut variables voltage u and current i). The rotor is modeled mechanically as a rigid body with a moment of inertia “J” and a viscous friction coefficient “b” accounting for the air drag and viscous friction in the lubricated bearings. The electric model of the rotor is given by the lumped parameters R and L, the rotor winding circuit from the electrical cut (u, i) towards the mechanical cut (T, ω).

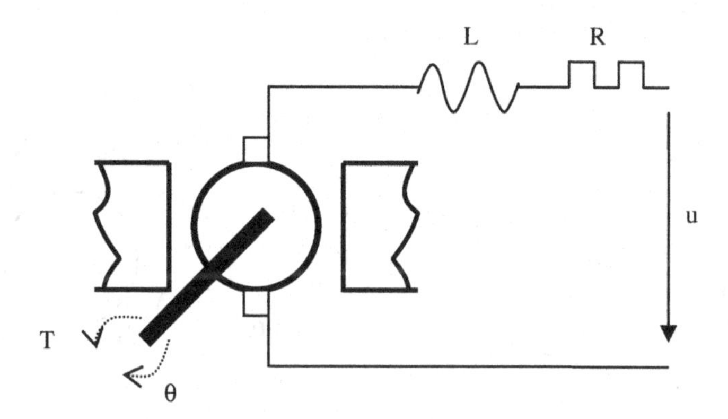

Fig. 1.8 The diagram of a PM- DC motor

The conversion of the electrical energy from the DC power source into the mechanical energy supplied to the load takes place in the DC motor, in particular in the electromagnetic field of the air gap between the stator and the rotor. Forces applied on rotor coils are generated as a result of the current i flowing through the rotor winding surrounded by the magnetic field produced by the PM of the stator. At the same time, the so called back electromotive force (back e.m.f.) are induced voltages in the moving rotor winding moving in the magnetic field. These two effects in a PM-DC motor can be modeled by separating the mechanical subsystem and the electrical subsystem, each being modeled by two port elements, as shown in Figs. 1.9 and 1.10, respectively.

In the left hand side of Fig. 1.9, torque components are represented around a cross section of the shaft. T_r denotes the torque generated in the electromagnetic field of the and acting on the rotor, while U_r represents the back electromagnetic force (back e.m.f.) induced by the magnetic field in the rotor winding in opposite to the supply voltage u. The torque T, and angular velocity ω are the cut variables toward the mechanical load, while the voltage u and the current i are the cut variables toward the DC power supply.

Example 1.5 Obtain the model equations.

The free body diagram and the two port circuit facilitate the derivation of the model equations.

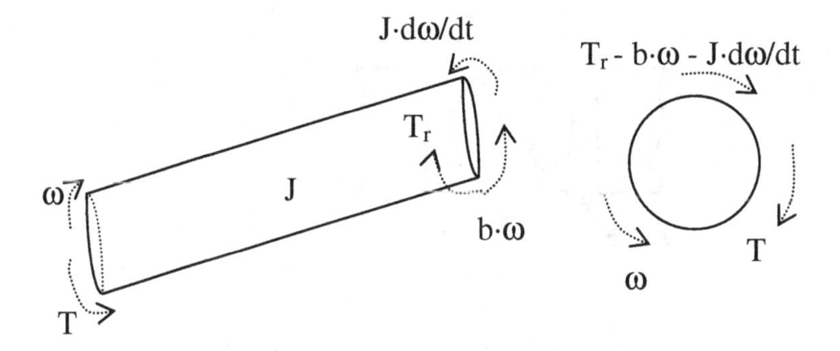

Fig. 1.9 Free body diagram for the mechanical part of the DC motor

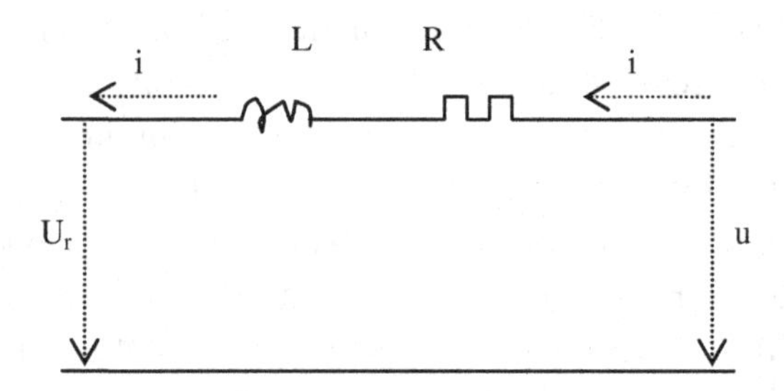

Fig. 1.10 Two port (U_r, i) and (u, i) circuit of the electrical part of the DC motor

Two algebraic equations result from the lumped-parameters model of the electro-mechanic conversion phenomena

$$T_r = k_m \cdot i$$

$$U_r = k_e \cdot \omega$$

where k_m [Nm/A] is the torque constant and k_e [Vs/rad] is the electrical constant.

In case of ideal conversion efficiency, $\eta = 1$, of the electrical power $U_r \cdot i$ into mechanical power $T_r \cdot \omega$,

$$\eta = (U_r \cdot i)/(T_r \cdot \omega) = 1$$

which gives

$$T_r \cdot \omega = U_r \cdot i$$

Using the above two algebraic equations, the following relationship is obtained

$$k_m \cdot i \cdot \omega = k_e \cdot \omega \cdot i$$

such that, in appropriate metric units, k_m in [Nm / A] and k_e in [Vs / rad], are of equal value

$$k_m = k_e$$

Power losses occur due to winding resistance, magnetic losses, friction etc. In the case of negligible losses, ideal power conversion can be assumed ($\eta = 1$).

For the mechanical part, shown in the free body diagram of Fig. 1.11, Newton second law gives:

$$J\frac{d\omega}{dt} = T_r - b \cdot \omega - T$$

For the electrical part shown in Fig. 1.12, the voltage drop equation gives:

$$u = L\frac{di}{dt} + R \cdot i + U_r$$

The last two differential equations and the two algebraic equations regarding the electro-mechanic conversion form a system of four differential-algebraic equations containing six variables T, ω, T_r, U_r, i and u. This system of four differential-algebraic equations represents the analytical model of the PM-DC motor.

The elimination of internal variables T_r and U_r results in a model reduced to two differential equations with four variables of the two cuts (T, ω) and (u, i):

$$k_m \cdot i = J\frac{d\omega}{dt} + b \cdot \omega + T$$

$$u = L\frac{di}{dt} + R \cdot i + k_e \cdot \omega$$

Most DC motors have negligible L, such that the model, for L = 0, is reduced to:

$$J\frac{d\omega}{dt} = k_m \cdot i - b \cdot \omega - T$$

$$u = R \cdot i - k_e\omega$$

These equations, obtained using effort-flow cuts, permit the determination of the electrical power $u \cdot i$ and mechanical power $T \cdot \omega$ transferred between these subsystems.

1.2.3 *Lagrange Equations for a Mixed Electro-Mechanical System*

Lagrange equations are given by [11]:

$$\frac{d}{dt}\frac{\partial}{\partial \dot{q}_r}[K] - \frac{\partial}{\partial q_r}[K] + \frac{\partial}{\partial q_r}[U] = Q_r \qquad \text{for } r = 1,2,\ldots..N$$

where

K is kinetic energy
U is potential energy
q_r is the generalized coordinate k
Q_r is the generalized force corresponding to the work done by the generalized coordinate q_r (or voltage in the case of the electrical generalized coordinate)
N is the total number of generalized coordinates needed to completely describe in time the components of the system.

For an electromechanical system with one generalized coordinate x for the mechanical part and one generalized coordinate Q for the electrical part, Lagrange equations for the mechanical and electrical parts of the system are given by [9, 11]:

$$\frac{d}{dt}\frac{\partial}{\partial \dot{x}}[K_m + K_e] - \frac{\partial}{\partial x}[K_m + K_e] + \frac{\partial}{\partial x}[U_m + U_e] = F$$

$$\frac{d}{dt}\frac{\partial}{\partial \dot{Q}}[K_m + K_e] - \frac{\partial}{\partial Q}[K_m + K_e] + \frac{\partial}{\partial Q}[U_m + U_e] = V$$

where

$K_m + K_e$ are the electric and mechanical kinetic energies
$U_m + U_e$ are the electric and mechanical potential energies
x is the generalized displacement variable (angular displacement)
$\dot{x} = v$ is the generalized velocity (angular velocity)
Q is the charge in capacitive components
$\dot{Q} = i$ is the current
F is the generalized force (dissipative and applied forces or torques)
V is the voltage (dissipative voltage drop and applied voltage)

Example 1.6 Obtain the model for the DC motor using Lagrange equations.

Lagrange equations for a PM- DC motor

For the DC motor shown in Fig. 1.8, Lagrange equations are

$$\frac{d}{dt}\frac{\partial}{\partial \dot{\theta}}[K_m + K_e] - \frac{\partial}{\partial \theta}[K_m + K_e] + \frac{\partial}{\partial \theta}[U_m + U_e] = F$$

$$\frac{d}{dt}\frac{\partial}{\partial \dot{Q}}[K_m + K_e] - \frac{\partial}{\partial Q}[K_m + K_e] + \frac{\partial}{\partial Q}[U_m + U_e] = V$$

or, taking into account that

$$\dot{\theta} = \omega$$

and

$$\dot{Q} = i$$

$$\frac{d}{dt}\frac{\partial}{\partial \omega}[K_m + K_e] - \frac{\partial}{\partial \theta}[K_m + K_e] + \frac{\partial}{\partial \theta}[U_m + U_e] = F$$

$$\frac{d}{dt}\frac{\partial}{\partial i}[K_m + K_e] - \frac{\partial}{\partial Q}[K_m + K_e] + \frac{\partial}{\partial Q}[U_m + U_e] = V$$

where

$$K_m(\omega) = J \cdot \omega^2/2$$

$$U_m = 0$$

$$F(\omega, i) = -b \cdot \omega + k_m \cdot i - T$$

$$K_e(i) = L \cdot i^2/2$$

$$U_e = 0$$

$$V(I, \omega) = u - R \cdot i - k_e \cdot \omega$$

Partial derivatives are

$$\frac{\partial}{\partial \omega}[K_m + K_e] = J\omega$$

$$\frac{\partial}{\partial \theta}[K_m + K_e] = 0$$

$$\frac{\partial}{\partial \theta}[U_m + U_e] = 0$$

$$\frac{\partial}{\partial i}[K_m + K_e] = Li$$

$$\frac{\partial}{\partial Q}[K_m + K_e] = 0$$

$$\frac{\partial}{\partial Q}[U_m + U_e] = 0$$

such that, for $k_m = k_e$, Lagrange equations result as follows

$$\frac{d}{dt}(J \cdot \omega) = k_m \cdot i - b \cdot \omega - T$$

$$\frac{d}{dt}(L \cdot i) = u - R \cdot i - k_e \cdot \omega$$

These are the same as the equations derived for the same DC motor using Effort-Flow representation of mixed systems and Newton-Euler equations of motion and Kirchhoff equations for electric circuits. Dissipative components are the dissipative voltage drop Ri and the dissipative generalized force, in this case the dissipative reaction torque $b \cdot \omega$.

Indeed, Lagrangian dynamics approach does not require effort-flow cuts neither for the mechanical subsystem nor for the electrical subsystem, and no internal variables were defined for such cases. For the interface between electrical and mechanical subsystems, applied torques T (external load torque) and $k_e \cdot i$ (motor torque) and applied voltages u (external voltage) and $k_e \cdot \omega$ (induced voltage) had to be however identified and this requires in fact the definition of the effort-flow cut at this interface.

Example 1.7 Figure 1.11 shows a plunger solenoid consisting of a solenoid of inductance L(x), dependent of the displacement x of the plunger from the non-energized position x = 0. The motion of the plunger along x is due to the plunger induced force, caused by the solenoid current i. The current flows in the electric circuit R-L(x) subject to the applied external voltage u(t). On the mechanical side, the plunger of mass M consists of a flexible rod with stiffness coefficient k supported by a lubricated bearing with viscous friction coefficient b. Obtain the model using Lagrange equations.

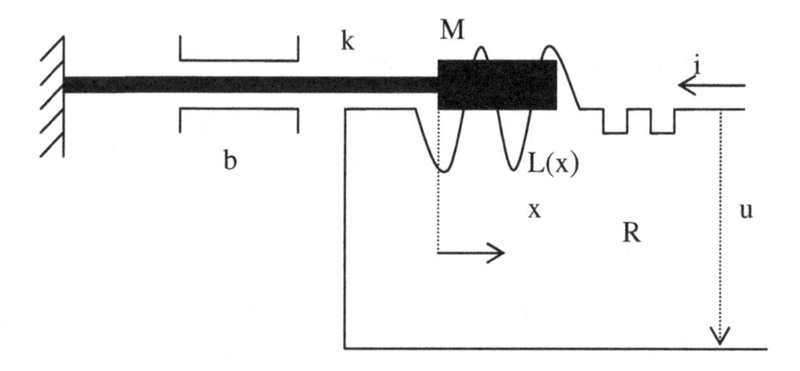

Fig. 1.11 The diagram of a plunger solenoid

Lagrange equations in this case are

$$\frac{d}{dt}\frac{\partial}{\partial \dot{x}}[K_m + K_e] - \frac{\partial}{\partial x}[K_m + K_e] + \frac{\partial}{\partial x}[U_m + U_e] = F$$

$$\frac{d}{dt}\frac{\partial}{\partial \dot{Q}}[K_m + K_e] - \frac{\partial}{\partial Q}[K_m + K_e] + \frac{\partial}{\partial Q}[U_m + U_e] = V$$

where

$$\dot{x} = v$$

and

$$\dot{Q} = i$$

such that

$$\frac{d}{dt}\frac{\partial}{\partial v}[K_m + K_e] - \frac{\partial}{\partial x}[K_m + K_e] + \frac{\partial}{\partial x}[U_m + U_e] = F$$

$$\frac{d}{dt}\frac{\partial}{\partial i}[K_m + K_e] - \frac{\partial}{\partial Q}[K_m + K_e] + \frac{\partial}{\partial Q}[U_m + U_e] = V$$

and

$$K_m(\omega) = M \cdot v^2/2$$

$$U_m = k \cdot x^2/2$$

$$F = -b \cdot v$$

$$K_e = L(x) \cdot i^2/2$$

$$U_e = 0$$

$$V = u(t) - R \cdot i$$

Partial derivatives are

$$\frac{\partial}{\partial v}[K_m + K_e] = M \cdot v$$

$$\frac{\partial}{\partial x}[K_m + K_e] = \frac{i^2}{2}\frac{d}{dx}L(x)$$

$$\frac{\partial}{\partial x}[U_m + U_e] = k \cdot x$$

$$\frac{\partial}{\partial i}[K_m + K_e] = L(x) \cdot i$$

$$\frac{\partial}{\partial Q}[K_m + K_e] = 0$$

$$\frac{\partial}{\partial Q}[U_m + U_e] = 0$$

such that Lagrange equations for the mechanical generalized coordinate x and for the electrical generalized coordinate Q result as follows

$$\frac{d}{dt}M \cdot v - \frac{i^2}{2}\frac{dL(x)}{dx} + k \cdot x = -b \cdot v$$

$$\frac{d}{dt}L(x) \cdot i = u(t) - R \cdot i$$

The term $(i^2 / 2) \cdot dL(x) / dx$ corresponds to the position dependent force applied by the solenoid on the plunger, while the term $d / dt\ L(x) \cdot i$ corresponds to the position dependent voltage drop on the solenoid inductance.

In reference [11] can be found other examples of Lagrangian dynamics for an electromechanical system in which there is a position dependent capacitance and for an angular position dependent mutual inductance.

In this section, the same equations of motion of an electro-mechanical system were obtained using two approaches effort-flow cuts with Newton-Kirchhoff dynamics and Lagrangian dynamics. The letter approach is particularly interesting due to the link to Hamiltonian dynamics and Lyapunov stability analysis for mixed systems [9].

Example 1.8 Figure 1.12 shows an electromechanical system composed of a spring, with spring coefficient k, and a coil of radius ρ, with moment of inertia J and with N turns in which flows a current $I = dQ/dt$ [11]. The angular position of the coil with regard to the horizontal plane is θ and varies from 0° to 180°. The coil is subject to a magnetic field produced by a solenoid with n turns in which flows a current $i = dq/dt$. The angular displacement of the coil is due to the induced torque resulting the solenoid current I and coil current i. Resistances of the coil and of the solenoid are R and r, respectively. The coil is subject to a voltage U(t) while the solenoid is subject voltage u(t). Self-inductances L of the coil and l of the solenoid are constant, i.e. independent of the angular position θ of the coil. The mutual inductance $M(\theta)$, between the static solenoid and the rotating coil, is dependent of the angular position θ of the coil

$$M(\theta) = k_{nN} \cdot \pi \cdot \rho^2 \cdot n \cdot N \cdot \sin\theta$$

where k_{nN} is a characteristic constant of the coil.

The coil is supported by a lubricated bearing with viscous friction coefficient B. Obtain the model using Lagrange equations.

Lagrange equations in this case are

$$\frac{d}{dt}\frac{\partial}{\partial\dot{\theta}}[K_m + K_e] - \frac{\partial}{\partial\vartheta}[K_m + K_e] + \frac{\partial}{\partial\theta}[U_m + U_e] = F$$

$$\frac{d}{dt}\frac{\partial}{\partial\dot{Q}}[K_m + K_e] - \frac{\partial}{\partial Q}[K_m + K_e] + \frac{\partial}{\partial Q}[U_m + U_e] = V$$

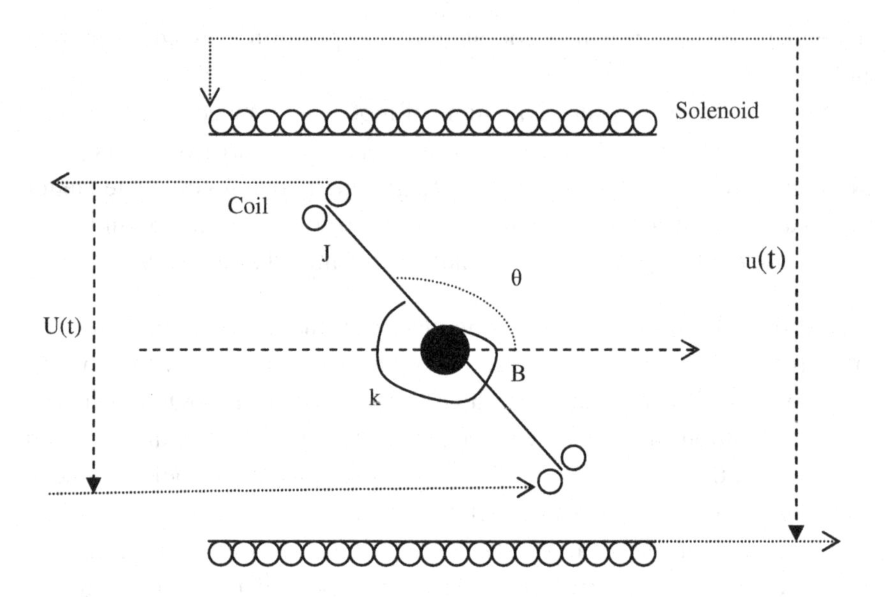

Fig. 1.12 The diagram of rotating spring coil and solenoid system

$$\frac{d}{dt}\frac{\partial}{\partial \dot{q}}[K_m + K_e] - \frac{\partial}{\partial q}[K_m + K_e] + \frac{\partial}{\partial q}[U_m + U_e] = v$$

where

$$\frac{d}{dt}\theta = \omega$$

$$\frac{d}{dt}Q = I$$

$$\frac{d}{dt}q = i$$

such that

$$\frac{d}{dt}\frac{\partial}{\partial \omega}[K_m + K_e] - \frac{\partial}{\partial \theta}[K_m + K_e] + \frac{\partial}{\partial \theta}[U_m + U_e] = F$$

$$\frac{d}{dt}\frac{\partial}{\partial I}[K_m + K_e] - \frac{\partial}{\partial Q}[K_m + K_e] + \frac{\partial}{\partial Q}[U_m + U_e] = V$$

$$\frac{d}{dt}\frac{\partial}{\partial i}[K_m + K_e] - \frac{\partial}{\partial q}[K_m + K_e] + \frac{\partial}{\partial q}[U_m + U_e] = v$$

and

$$K_m(\omega) = J\frac{\omega^2}{2}$$

$$U_m = k\frac{\theta^2}{2}$$

$$F = -B \cdot \omega$$

$$K_e = L\frac{I^2}{2} + l\frac{i^2}{2} + M(\theta) \cdot I \cdot i = L\frac{I^2}{2} + l\frac{i^2}{2} + k_{nN} \cdot \pi \cdot \rho^2 \cdot n \cdot N \cdot \sin(\theta) \cdot I \cdot i$$

or

$$K_e = L\frac{I^2}{2} + l\frac{i^2}{2} + \alpha \cdot I \cdot i \cdot \sin\theta$$

$$U_e = 0$$

$$V = U(t) - R\,I$$

$$v = u(t) - r\,i$$

where the constant α is

$$\alpha = k_{nN} \cdot \pi \cdot \rho^2 \cdot n \cdot N$$

Partial derivatives are

$$\frac{\partial}{\partial\theta}[K_m + K_e] = \frac{\partial}{\partial\theta}\alpha \cdot I \cdot i \cdot \sin\theta = \alpha \cdot I \cdot i \cdot \cos\theta$$

$$\frac{\partial}{\partial Q}[K_m + K_e] = 0$$

$$\frac{\partial}{\partial q}[K_m + K_e] = 0$$

$$\frac{\partial}{\partial\theta}[U_m + U_e] = k \cdot \theta$$

$$\frac{\partial}{\partial I}[K_m + K_e] = L \cdot I + \alpha \cdot i \cdot \sin\theta$$

$$\frac{\partial}{\partial i}[K_m + K_e] = l \cdot i + \alpha \cdot I \cdot \sin\theta$$

$$\frac{\partial}{\partial Q}[K_m + K_e] = 0$$

$$\frac{\partial}{\partial q}[K_m + K_e] = 0$$

$$\frac{\partial}{\partial Q}[U_m + U_e] = 0$$

$$\frac{d}{dt}\frac{\partial}{\partial\omega}[K_m + K_e] = I \cdot \dot{\omega}$$

$$\frac{d}{dt}\frac{\partial}{\partial I}[K_m + K_e] = L \cdot \dot{I} + \alpha \cdot \dot{i} \cdot \sin\theta + \alpha \cdot i \cdot \omega \cdot \cos\theta$$

$$\frac{d}{dt}\frac{\partial}{\partial i}[K_m + K_e] = l \cdot \dot{i} + \alpha \cdot \dot{I} \cdot \sin\theta + \alpha \cdot I \cdot \omega \cdot \cos\theta$$

such that Lagrange equations for the mechanical generalized coordinate θ and for the electrical generalized coordinates Q and q result as follows

$$J \cdot \dot{\omega} - \alpha \cdot I \cdot i \cdot \cos\theta - k \cdot \theta = - B \cdot \omega$$

$$L \cdot \dot{I} + \alpha \cdot \dot{i} \cdot \sin\theta + \alpha \cdot i \cdot \omega \cdot \cos\theta = U(t) - R \cdot I$$

$$L \cdot \dot{i} + \alpha \cdot \dot{I} \cdot \sin\theta + \alpha \cdot I \cdot \omega \cdot \cos\theta = u(t) - r \cdot i$$

or

$$J \cdot \ddot{\theta} - B \cdot \dot{\theta} - k \cdot \theta = \alpha \cdot I \cdot i \cdot \cos\theta$$

$$L \cdot \dot{I} + R \cdot I = U(t) - \alpha \cdot \dot{i} \cdot \sin\theta - \alpha \cdot i \cdot \omega \cdot \cos\theta$$

$$L \cdot \dot{i} + r \cdot i = u(t) - \alpha \cdot \dot{I} \cdot \sin\theta - \alpha \cdot I \cdot \omega \cdot \cos\theta$$

These three nonlinear differential equations with variables θ(t), I(t) and i(t) represent the model of the system from Fig. 1.12, given the inputs U(t) and u(t), *i.e.* the direct problem. In practical applications, one of the inputs, U(t) or u(t), can be held constant. For either I or I vanishing, the first equation gives the equilibrium position θ = 0.

In the first equation, for the rotational mechanical subsystem, the term $T = \alpha \cdot I \cdot i \cdot \cos\theta$ represents the torque produced by the magnetic fields interaction of the solenoid with the coil, which is zero when the coil and the solenoid are perpendicular, i.e. when θ = 90°, or when the two magnetic fields are parallel. As a result, the angle θ should be limited to the domain

$$-90 + \varepsilon < \theta < 90 - \varepsilon$$

where ε can be obtained from the condition that maximum admissible currents I_{max} and i_{max} produce a minimum required torque T_{min} to be able to rotate the J-B-K mechanical system, i.e.

$$T_{min} = \alpha\, I_{max}\, i_{max} \cos|\theta - \varepsilon|$$

In second equation, for the moving coil, the terms $\alpha \cdot \dot{i} \cdot \sin\theta + \alpha \cdot i \cdot \omega \cdot \cos\theta$ represent the induced voltages in the coil due to the time varying current and due to the coil angular velocity. Similarly, the terms $\alpha \cdot \dot{I} \cdot \sin\theta + \alpha \cdot I \cdot \omega \cdot \cos\theta$ represent induced voltages in the solenoid due to the time varying current and due to the coil angular velocity.

Example 1.9 Figure 1.13 shows a capacitance with a moving top electrode of mass m and with a gap X – x, where X is the gap. The equilibrium position of the top electrode is x = 0, when no voltage is applied to the capacitance and the spring is stretched by m · g / k to counterbalance top electrode weight m · g. The bottom electrode is sitting on a fixed electric insulator. The top electrode can move vertically with the displacement x, as a result of the time varying voltage applied to the electrodes from a voltage source with U(t) connected through wires with resistance R and inductance L [11]. The top electrode is connected to the moving bottom end of a spring with spring coefficient k. The spring has the top end connected to a fixed insulator. Assume that the structural damping coefficient is b.

The capacity of the time varying gap capacitance is given by

$$C(x) = \frac{k \cdot A}{X - x}$$

where k is the dielectric constant and A is the cross-sectional area of the capacitance. Obtain the model using Lagrange equations.

Lagrange equations in this case are

$$\frac{d}{dt}\frac{\partial}{\partial \dot{x}}[K_m + K_e] - \frac{\partial}{\partial x}[K_m + K_e] + \frac{\partial}{\partial x}[U_m + U_e] = F$$

$$\frac{d}{dt}\frac{\partial}{\partial \dot{Q}}[K_m + K_e] - \frac{\partial}{\partial Q}[K_m + K_e] + \frac{\partial}{\partial Q}[U_m + U_e] = V$$

where

$$\frac{dx}{dt} = v$$

$$\frac{dQ}{dt} = I$$

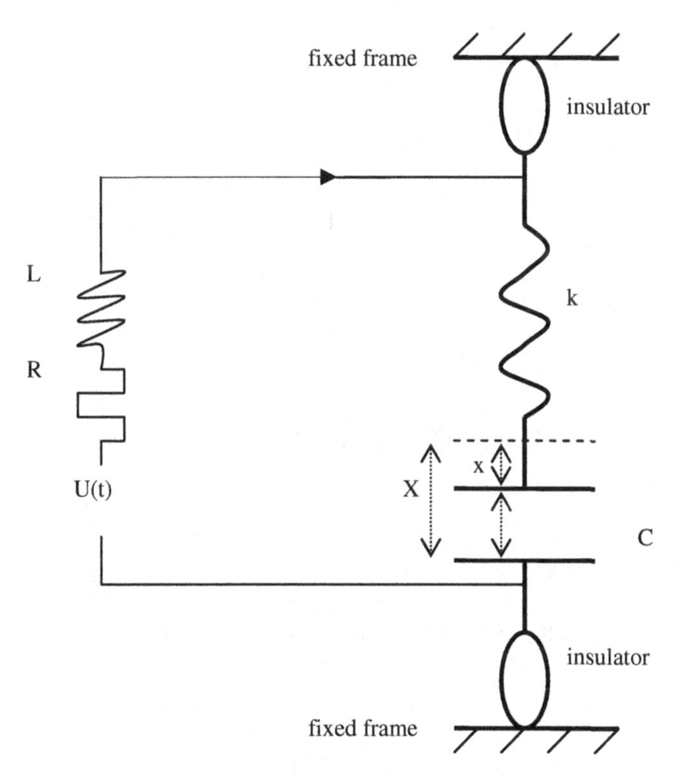

Fig. 1.13 Diagram of a system consisting of a capacitance and a spring

such that

$$\frac{d}{dt}\frac{\partial}{\partial v}[K_m + K_e] - \frac{\partial}{\partial x}[K_m + K_e] + \frac{\partial}{\partial x}[U_m + U_e] = F$$

$$\frac{d}{dt}\frac{\partial}{\partial I}[K_m + K_e] - \frac{\partial}{\partial Q}[K_m + K_e] + \frac{\partial}{\partial Q}[U_m + U_e] = V$$

and

$$K_m(v) = m\frac{v^2}{2}$$

$$U_m(x) = k\frac{x^2}{2}$$

$$F = -b \cdot v$$

$$K_e(I) = L\frac{I^2}{2}$$

$$U_e(x) = Q^2 \frac{X - x}{2 \cdot A \cdot k}$$

$$U_e(x) = Q^2 \cdot (X - x)/(2 \cdot A \cdot c)$$

Partial derivatives are

$$\frac{\partial}{\partial x}[K_m + K_e] = \frac{\partial}{\partial x}(m\frac{v^2}{2} + L\frac{I^2}{2}) = 0$$

$$\frac{\partial}{\partial Q}[K_m + K_e] = 0$$

$$\frac{\partial}{\partial x}[U_m + U_e] = \frac{\partial}{\partial x}[k\frac{x^2}{2} + Q^2 \cdot \frac{(X - x)}{2 \cdot A \cdot k}] = k \cdot x - \frac{Q^2}{2 \cdot A \cdot k}$$

$$\frac{\partial}{\partial v}[K_m + K_e] = \frac{\partial}{\partial v}(m\frac{v^2}{2} + L\frac{I^2}{2}) = m \cdot v$$

$$\frac{\partial}{\partial I}[K_m + K_e] = \frac{\partial}{\partial I}(m\frac{v^2}{2} + L\frac{I^2}{2}) = L \cdot I$$

$$\frac{\partial}{\partial Q}[K_m + K_e] = 0$$

$$\frac{\partial}{\partial Q}[U_m + U_e] = \frac{\partial}{\partial Q}[kx^2/2 + Q^2\frac{X-x}{2 \cdot A \cdot k}] = Q\frac{X-x}{A \cdot k}$$

$$\frac{d}{dt}\frac{\partial}{\partial v}[K_m + K_e] = m \cdot \dot{v}$$

$$\frac{d}{dt}\frac{\partial}{\partial I}[K_m + K_e] = L \cdot \dot{I}$$

Lagrange equations for the mechanical generalized coordinate θ and for the electrical generalized coordinates Q and q result as follows

$$m \cdot \dot{v} + k \cdot x - \frac{Q^2}{2 \cdot A \cdot k} = -b \cdot v$$

$$L \cdot \dot{I} + Q\frac{X - x}{A \cdot k} = U(t) - R \cdot I$$

The following two second order nonlinear differential equations with unknowns x(t) and Q(t) and q(t) represent the model of the system from Fig. 1.13, given the time varying input voltage U(t).

$$m \cdot \ddot{x} - b \cdot \dot{x} - k \cdot x = F(Q)$$

$$L \cdot \ddot{Q} + R \cdot \dot{Q} + \frac{Q}{C(x)} = U(t)$$

where C(x) is the time varying gap dependant capacitance with

$$C(x) = \frac{k \cdot A}{X - x}$$

and

$$F(Q) = \frac{Q^2}{2 \cdot k \cdot A}$$

is the charge dependant force applied by the moving electrode to the bottom end of the spring.

The two nonlinear differential equations with variables x and Q permit to model the effect of time varying external voltage U(t) on the displacement x(t) of the moving top electrode, i.e. a direct problem.

1.3 Local Sensing and Actuation in Spatially Continuous Systems

Spatially continuous systems, can be modeled using either effort-flow cuts or Lagrangian dynamics. These models are needed for the design of systems or for their real-time monitoring and control.

Continuous systems can be modeled with lumped parameters models or with distributed parameters models, depending on the acceptable level of accuracy and modeling difficulties. In both cases, the number of inputs can be lower than the number of degrees of freedom, resulting in under-actuation or lower number of outputs than states, resulting in under-sensing. The issue of local sensing and actuation has to be investigated in both cases. Control of these systems can be either open loop or closed loop. Under-actuation and under-sensing have consequences on the performance of both types of systems, but is a particularly difficult problem to solve for distributed parameters models [18].

1.3.1 *Lumped Parameters Models with Under-Actuation and Under-Sensing*

Lumped parameters models for linear case can be written in the form of linear ordinary differential equations (ODE):

$$d\mathbf{X(t)} / dt = \mathbf{A}(t) \cdot \mathbf{X(t)} + \mathbf{B}(t) \cdot \mathbf{u}(t) + \mathbf{G}(t) \cdot \mathbf{w}(t)$$

$$\mathbf{y}(t) = \mathbf{C}(t) \cdot \mathbf{X(t)} + \mathbf{D}(t) \cdot \mathbf{u}(t)$$

where

$\mathbf{X(t)}$ = n-vector of states with given initial conditions $\mathbf{x}(0)$
$\mathbf{u}(t)$ = m-vector of inputs
$\mathbf{w}(t)$ = d-vector of disturbances
$\mathbf{y}(t)$ = p-vector of outputs
$\mathbf{A}(t)$, $\mathbf{B}(t)$, $\mathbf{G}(t)$, $\mathbf{C}(t)$, $\mathbf{D}(t)$ = time varying matrices.

Lumped models for nonlinear case can also be written in the form of linear ordinary differential equations (ODE):

$$d\mathbf{X(t)} / dt = \mathbf{F}(\mathbf{X(t)}, \mathbf{u}(t), \mathbf{w}(t))$$

$$\mathbf{y}(t) = \mathbf{H}(\mathbf{X(t)}, \mathbf{u}(t))$$

where F and H are nonlinear functions.

The number of states, n, is finite and, consequently, lumped parameters models which are a simplified representation of continuous systems. Certainly, spatial resolution is in the former case limited. Under-actuation results from fewer inputs m than the number of degrees of freedom N, i.e. $m < N$, and under-sensing from fewer outputs p than the number of states, i.e. $p < n$. A continuous system would have infinite values for n and N, consequently, finite number of actuators and sensors will always result in this case in under-actuation and under-sensing. Given the complexities of distributed parameters models, under-actuation and under-sensing issues are easier to be analyzed using in a first approximation lumped linear models represented by ordinary differential equations (ODE) with time invariant (LTI) parameters.

1.3.2 *Distributed Parameters Models with Under-Actuation and Under-Sensing*

Distributed parameters models can be take a large variety of mathematical forms. A generic form is:

$$\delta \mathbf{X}(x, y, z, t) / \delta t = \mathbf{F}(\mathbf{X}(x, y, z, t), \nabla \mathbf{X}(x, y, z, t), \nabla^2 \mathbf{X}(x, y, z, t), \ldots, \mathbf{w}(t))$$

subject to boundary conditions

$$\mathbf{G}(\mathbf{X}(x_b, y_b, z_b, t), \mathbf{u}(t)) = 0$$

and intial conditions

$$\mathbf{I}(\mathbf{X}(x, y, z, 0), \mathbf{u}(0)) = 0$$

While, output equation is

$$\mathbf{H}(\mathbf{y}(x_m, y_m, z_m, t), \mathbf{X}(x, y, z, t), \mathbf{u}(t))$$

where ∇ is the partial differentiation operator, with regard to x, y, z, variables and the function G and the subscript b refer to boundary conditions, while the function I defines initial conditions. It can be observed that control variables $\mathbf{u}(t)$ appear in this case only in the boundary conditions, a typical case in practice where the continuous system is actuated only from specific system boundaries. Similarly, the outputs $\mathbf{y}$ are typically measured in some specific points x_m, y_m, z_m. These limitations regarding local actuation and sensing pose specific challenges to the design and performance of controllers and for the integration of spatially continuous systems.

1.4 Centralized versus Local Control

Local sensing and actuation of systems with large or infinite number of states is linked also to the issue of centralized versus local control. A finite number of actuators can be controlled either at the actuator location or using a centralized control for all actuators.

Local controllers use collocated actuators and sensors, have the advantage of easier design and tuning and tend to produce predictable local system behavior, but are not optimal for the system as a whole. Moreover, dynamic couplings in the system can result in inefficient or unstable system behavior. Centralized control can be designed optimally, but suffers from unavoidable simplifications of the system model on which they are based and requires often a prohibitively large number of signal transmissions [19]. These issues are critical for continuous systems distributed over a large area or for formations.

Problems

1. Consider the system shown in Fig. 1.2 but with added viscous friction between the mass M and the ground, with viscous friction coefficient B. Obtain v(s) given F(s).

2. For the system shown in Fig. 1.5, obtain the four cuts representation.

3. For the free body diagram shown in Fig. 1.7, consider that the mass of the rod is not negligible and that is concentrated equally at the two ends of the diagram as M_1 and M_2. Obtain the equations for v_1 and F_1 function of v_2 and F_2.

4. For the DC motor shown in Fig.1.8, assume that the shaft is flexible, such that in the free body diagram from Fig. 1.9 a torsional spring coefficient K is in series with the moment of inertia J.
   a. Obtain the model with two differential equations for the cut variables (T, ω) and (u, i)
   b. Verify that the same model is obtained using Lagrange equations.

5. Assume that the plunger solenoid from Fig. 1.11 has the plunger of mass M connected by a spring, with spring coefficient K, to a

right hand side rigid wall. Obtain the Lagrange equations of motion.

6. For electromechanical system shown in Fig. 1.12, the mutual inductance between the static solenoid and the rotating coil is $M(\theta) = k_{nN} \cdot \pi \cdot \rho^2 \cdot n \cdot N \cdot \sin\theta$. The coil, of moment of inertia J, actuates a flexible shaft supported at one end by a lubricated bearing with viscous friction coefficient B. The shaft, with torsional stiffness coefficient K, has a load with a moment of inertia J, and has itself a negligible moment of inertia, relative to the two end moments of inertia. Obtain Lagrange equations for this system.

7. Consider the system shown in Fig. 1.13, which consists of a capacitance with a moving top electrode of mass m and with a gap X - x, where X is the gap for the equilibrium position x = 0, when no voltage is applied to the capacitance, and the spring is stretched by $m \cdot g / k$ to counterbalance top electrode weight $m \cdot g$. The bottom electrode is sitting on a fixed insulator. The top electrode is moving vertically with the displacement x, as a result of the time varying voltage applied to the electrodes from a voltage source with U(t) connected through wires with resistance R and inductance L. The top electrode is connected to the moving bottom end of a spring with spring coefficient k and in parallel with a damper with damping coefficient b. The spring has the top end connected to a fixed insulator. The capacity of the time varying gap capacitance is $C(x) = c \cdot A / (X - x)$, where c is a constant dependent of the insulator between the electrodes. Obtain the model using Lagrange equations.

8. For a multi-DOF linear lumped parameters mechanical system, the system is considered under-actuated if:
   a. there are fewer actuators than the number of states
   b. there are as many actuators as the number of states
   c. there are as many actuators as the number of degrees of freedom.

Chapter 2

Examples of Direct and Inverse Problems for Mixed Systems

2.1 Modular Modeling and Control Issues for Mixed Systems

2.1.1 *Effort-Flow Modeling of Mechatronic Systems*

Issues regarding direct and inverse problems can be presented based on the schematic diagram of a distributed parameters mechatronic system, shown in Fig. 2.1. In Fig. 2.1, measurement and control signal transmission is shown in thin lines, while power transfer to the Distributed Parameters Mixed System is represented using effort-flow representation of power transfer cuts concept, introduced in Ch. 1.

Electric power supply of the mechatronic system provides an instantaneous power transfer $U_{AC} \cdot I_{AC}$ to supply the drivers. The drivers modulate the electric power output $u \cdot i$ to actuators, assumed electric motors with given efficiency, such that $u \cdot i < U_{AC} \cdot I_{AC}$. Actuators can be controlled by modulating voltage or current input. The modulation follows the computer control commands transmitted from the DAC as analog signals. The actuators, assumed here as electromechanical actuators, provide modulated mechanical power $Fv < ui < U_{AC} \cdot I_{AC}$ to the Distributed Parameters Mixed System for changing, as required, the states of the system. In Fig. 2.1, point actuators are assumed to apply forces F in given points at the outer boundary surface of the system.

Inside the Distributed Parameters Mixed System other power transfer and conversions take place and they can also be represented by the effort-flow representation of power transfer cuts concept in case that this is concentrated in specific some points of the system. Distributed parameters power transfer and conversion, for example in case of radiation, require specific distributed parameters effort-flow representation in which power transfer cross-sections can be identified [23].

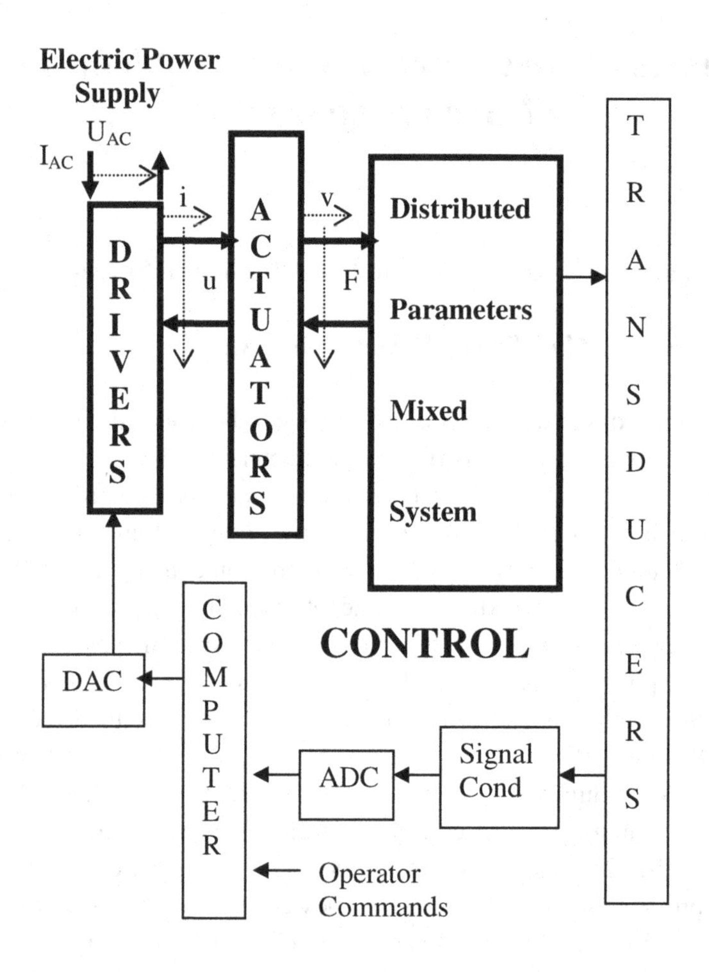

Fig. 2.1 Schematic diagram of a distributed parameters mechatronic system with effort-flow representation of power transfer cuts

Modeling and simulation of Distributed Parameters Mixed Systems has to be based on the fact that, in this case states, inputs and outputs are dependent not only on time but also on spatial x, y, z location in the system.

2.2 Modeling and Simulation of Distributed Parameters Systems

2.2.1 *Examples of Distributed Parameters Systems*

2.2.1.1 *Examples of models of vibrating flexible structures*

Flexible structures (strings, membranes, beams, plates etc.), acoustic field, heat transfer, fluid flow, electric and magnetic fields, are some examples of systems that have distributed parameters and are modeled by partial differential equations or alternatively, by integral equations [14, 24]. While the initial model of such systems is in the form of distributed parameters, often, for developing active control of the dynamics of these systems, an equivalent lumped parameters model is often derived, as for example the finite elements model for vibrating systems [23].

A) Examples of models of vibrating flexible structures are the following:

a) the string, shown in Fig. 2.2, has a small transversal displacement y(x, t) from the equilibrium position. In this example, initial conditions, away from equilibrium lead to space and time variation of y(x, t).

The motion equation is

$$\frac{\partial^2 y(x,t)}{\partial t^2} = c^2 \frac{\partial^2 y(x,t)}{\partial x^2}$$

where

$c = T/\rho$ [m / s]

T is the constant tension in the string [N]

ρ is linear mass density [kg / m]

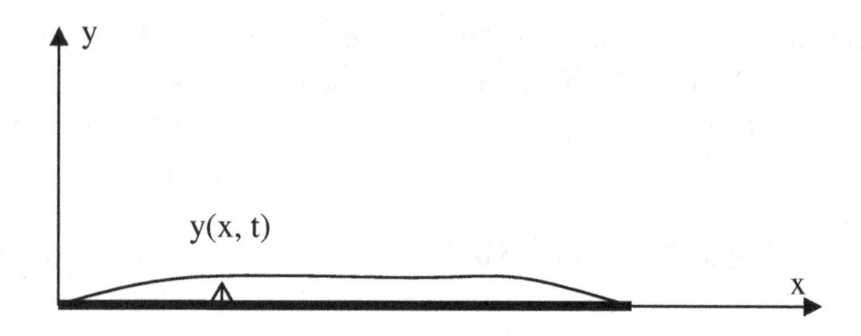

Fig. 2.2 Vibrating string

b) the membrane, shown in Fig. 2.3 has a small transversal displacement z(x, y, t) from the equilibrium position.

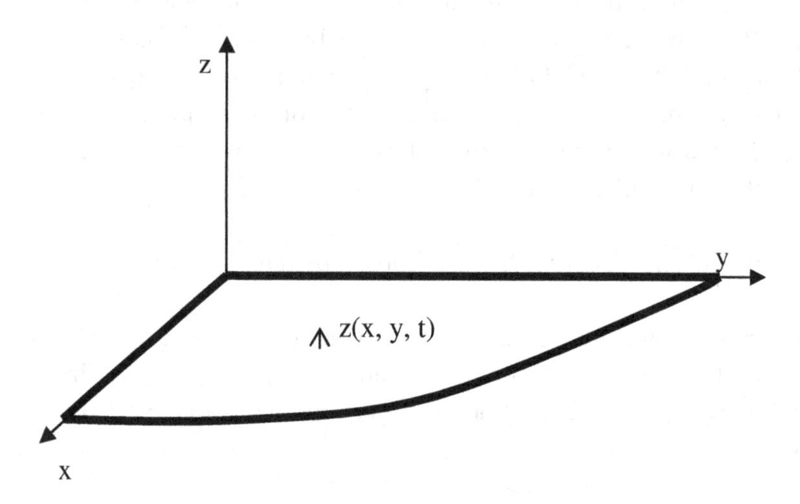

Fig. 2.3 Vibrating membrane

As in the case of vibrating string, the displacement is governed by second order partial differential equation with regard to time and to space coordinates.

The equation is

$$\frac{\partial^2 z(x,y,t)}{\partial t^2} = c^2 \left(\frac{\partial^2 z(x,y,t)}{\partial x^2} + \frac{\partial z^2(x,y,t)}{\partial y^2}\right)$$

where c = T/ρ [m / s]. Initial conditions are assumed away from equilibrium.

c) the beam is shown in Fig. 2.4 and has a small longitudinal displacement u(x, t), along axis x, from the equilibrium position.

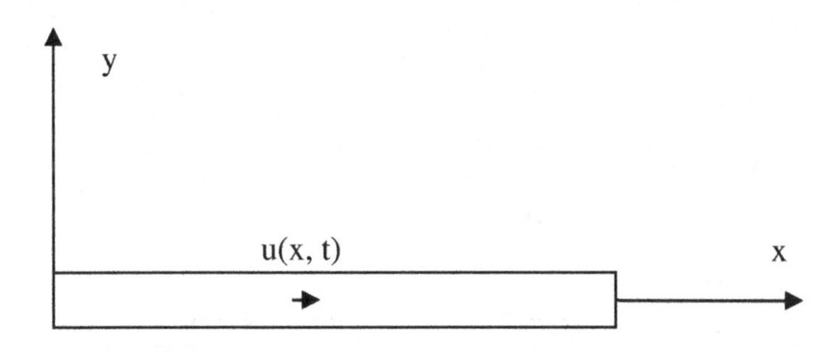

Fig. 2.4 Beam vibrating longitudinally

The equation is the same the equation for vibrating string, but, written for the longitudinal displacement u(x, t)

$$\frac{\partial^2 u(x,t)}{\partial t^2} = c^2 \frac{\partial^2 u(x,t)}{\partial x^2}$$

where
$c = T / \rho$ [m / s]
T is the constant tension in the beam [N]
ρ is linear mass density [kg / m]

d) the beam, shown in Fig. 2.5, has a small transversal displacement y(x, t) from the equilibrium position and is subject to an applied distribute force F(x, t).

The equation is

$$\frac{\partial^2 y(x,t)}{\partial t^2} + b^2 \frac{\partial^4 y(x,t)}{\partial x^4} = b^2 \frac{F(x,t)}{E \cdot I}$$

where

$b = E \cdot I \cdot g / \mu$ [m / s]
E is the Young modulus of the homogenous material of the beam
I is the moment of inertia about x axis
μ is linear mass density [kg / m]

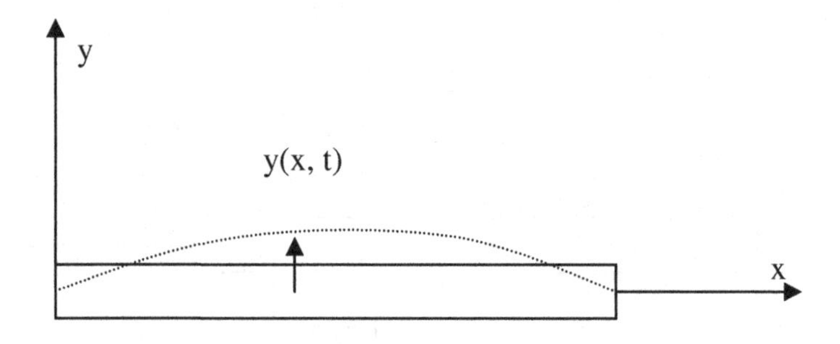

Fig. 2.5 Beam vibrating transversally

Different from the previous examples of flexible structures, which had vibrations due to initial nonzero conditions, in this case there is an external force F(x, t) applied to the beam.

2.2.1.2 *Acoustic fields*

Acoustic fields of relatively low amplitudes are modeled by the linear Euler equation [21]

$$\frac{\partial^2 p}{\partial t^2} = c^2 \cdot \nabla^2 p$$

where
p(x, y, z, t) is the acoustic pressure at x, y, z at time t
c is the thermodynamic speed of sound for the specific fluid supporting the propagation.

∇^2, nabla squared, is the Laplacian operator defined by

$$\nabla^2 = \frac{\partial^2}{\partial x^2} + \frac{\partial^2}{\partial y^2} + \frac{\partial^2}{\partial z^2}$$

2.2.1.3 *Heat transfer*

Three-dimensional (3D) heat conduction equation is given by

$$\frac{\partial u}{\partial t} = k \cdot \nabla^2 u$$

where
u(x, y, z, t) is the temperature in a solid body in the point x, y, z at time t.
k is diffusivity.
σ is the specific heat of the solid body conducting the heat
τ is the volume density [kg / m^3]

2.2.1.4 *Fluid flow*

Euler's method for the flow in space gives the state change f(x, y, z, t) of a particle of the fluid as follows

$$\frac{df}{dt} = \frac{\partial f}{\partial t} + V \cdot \text{grad}(f)$$

where
V is the velocity
and the gradient operator grad (nabla or del) is given by

$$\text{grad} = \nabla = i\frac{\partial}{\partial x} + j\frac{\partial}{\partial y} + k\frac{\partial}{\partial z}$$

2.2.1.5 *Electric and magnetic fields*

For an infinite, homogenous, isotropic, insulating (I = 0), *i.e.* no free charges (Q = 0), lossless, dispersion-less and linear medium, i.e. an ideal vacuum free space, Maxwell equations for electromagnetic fields can be written in the partial differential equations form, in Cartesian coordinates as follows

-Gauss equation for electric field intensity **E**

$$\delta E_x / \delta x + \delta E_y / \delta y + \delta E_z / \delta z = 0$$

-Gauss equation for magnetic induction **B**

$$\delta B_x / \delta x + \delta B_y / \delta y + \delta B_z / \delta z = 0$$

-Faraday law

$$\begin{aligned} \delta E_z / \delta y - \delta E_y / \delta z &= - \delta B_x / \delta t \\ \delta E_x / \delta z - \delta E_z / \delta x &= - \delta B_y / \delta t \\ \delta E_y / \delta x - \delta E_x / \delta y &= - \delta B_z / \delta t \end{aligned}$$

-Ampere law

$$\begin{aligned} \delta B_z / \delta y - \delta B_y / \delta z &= \mu \cdot A \cdot \varepsilon \cdot \delta E_x / \delta t \\ \delta B_x / \delta z - \delta B_z / \delta x &= \mu \cdot A \cdot \varepsilon \cdot \delta E_y / \delta t \\ \delta B_y / \delta x - \delta B_x / \delta y &= \mu \cdot A \cdot \varepsilon \cdot \delta E_z / \delta t \end{aligned}$$

where
E = electric field with the components E_x, E_y E_z [V / m]
B = magnetic flux density or magnetic induction, [T] or [N / (Am)], with the components B_x , B_y , B_z
μ = permeability constant = $\mu_r \cdot \mu_o$, [H / m] or [N / A^2]
μ_r = relative permeability constant with regard to vacuum ($\mu_r \approx 1$ for air)
μ_o = vacuum permeability constant = $4 \cdot \pi \cdot 10^{-7}$ [H / m] or [N / A^2]
ε = permittivity constant [F/m] or [C^2 / Nm^2]
ε_r = relative permittivity constant with regard to vacuum, ($\varepsilon_r \approx 1$ for air)

ε_0 = vacuum permittivity constant = $8.85 \cdot 10^{-12}$ [F / m] or [C^2 / Nm^2]

The above examples of distributed parameters models, expressed as partial differential equations, require boundary and initial conditions for complete definition for a specific system.

These systems can have external excitations from actuators, that are located on some points along the 1D, 2D or 3D field. For 3D fields in particular, actuators are often located on the 2D outer boundary surface of the field and appear in the model only in the boundary conditions. Sometimes, as shown in the example shown in Fig. 2.5, actuating force can be applied distributed within the system. The placement of actuators in distributed parameters systems has important consequences on the design of controllers for both open loop and closed loop control.

These problems will be presented in more detail in subsequent chapters.

2.2.2 *Direct and Inverse Problems. Well Posed and Ill Posed Problems*

Two distinct problems can be formulated for distributed parameters systems:
-simulation problem, to determine positions and time dependent states and outputs given distributed external excitation and initial and boundary conditions. This is called a direct problem: given known input determine the output.
-control problem, to determine distributed external excitation, applied often on the boundary, given desired states and outputs as well as initial and boundary conditions. This is an inverse problem: given desired outputs determine inputs.

These problems can be characterized as well-posed if [82]:
-the solution exists;
-the solution is unique;
-the model is continuously dependent on parameters.

These conditions, stated by Hadamard, when violated define ill-posed problems. In particular, partial differential equations are characterized as ill-posed. Ch.3 presents methods for solving discrete inverse problems for the case of ill-posedness.

2.2.3 *Classification of Partial Differential Equations and Methods of Solving*

The majority of partial differential equations (PDE) used in mathematical physics are of the following types:
-Single linear PDE with one unknown, for example for heat flow, string equation, Euler Bernoulli beam equation and membrane equation as well as wave equation for vibrations, sound and electromagnetic waves;
-Multiple linear PDE with multiple unknowns, for example Maxwell equations;
-Nonlinear systems of PDE, for example Euler and Navier-Stokes equations for fluid dynamics.

Most of these PDE in mathematical physics are second order equations and, in case of two independent variables, are of the general form:

$$A\frac{\partial^2 u(x,y)}{\partial x^2}+B\frac{\partial^2 u(x,y)}{\partial x\partial y}+C\frac{\partial^2 u(x,y)}{\partial y^2}+D\frac{\partial u(x,y)}{\partial x}+E\frac{\partial u(x,y)}{\partial y}$$
$$+Fu(x,y)=G$$

or, in a more compact notation

$$A\cdot u_{xx}+B\cdot u_{xy}+C\cdot u_{yy}+D\cdot u_x+E\cdot u_y+F\cdot u=G$$

where A, B, C, D, E, F, G do not depend on u but might depend on x and y.

Second order equations PDE are classified based on the sign of $B^2-4\cdot A\cdot C$:

a) for $B^2-4\cdot A\cdot C<0$, elliptic equations.
For example two-dimensional heat conduction equation

$$\frac{\partial u(x,y,t)}{\partial t}=k(\frac{\partial^2 u(x,y,t)}{\partial x^2}+\frac{\partial^2 u(x,y,t)}{\partial y^2})$$

in steady state form ($\delta u(x, y, t) / \delta t = 0$ for $t = \infty$) becomes Laplace equation with independent variables x, y

$$\frac{\partial^2 u(x,y,\infty)}{\partial x^2}+\frac{\partial^2 u(x,y,\infty)}{\partial y^2}=0$$

where $A = C = 1$ and $B = 0$ such that $B^2 - 4 \cdot A \cdot C = -4 < 0$.

b) for $B^2 - 4 A \cdot C > 0$, hyperbolic equations.
For example string and longitudinal vibrations equations with independent variables x and t

$$c^2 \frac{\partial^2 u(x,t)}{\partial x^2}-\frac{\partial^2 u(x,t)}{\partial t^2}=0$$

where $A = c^2$, $B = 0$, $C = -1$ and $B^2 - 4 \cdot A \cdot C = 4 \cdot C^2 > 0$.

c) for $B^2 - 4 \cdot A \cdot C = 0$, parabolic equations.
For example one-dimensional heat conduction equation

$$\frac{\partial u(x,t)}{\partial t}=k\frac{\partial^2 u(x,t)}{\partial x^2}$$

where $A = k$, $B = C = 0$ and $B^2 - 4 \cdot A \cdot C = 0$.

The importance of this classification is due to the fact that each class shares similar methods of solving the equations of the direct problem.

Methods for solving the above equations given initial and boundary conditions include [14, 24]:

A) analytical methods
-general and particular solutions
-separation of variables and modal analysis, often used in vibration engineering
-Fourier transform
-Laplace transform, in particular with regard to time variable etc.
B) numerical methods
-finite differences method
-finite elements method.

These methods will be reviewed in examples as part of the presentation of various applications in next chapters.

2.3 Overview of Open Loop and Closed Loop Control of Distributed Parameters Systems

2.3.1 *Direct and Inverse Problems*

Both open and closed loop control of systems use a model of the system for the controller design, to determine the control scheme that provides the commands for the inputs **u**(t) such that the states **X**(t) of the system tend towards some given desired values or time variations. System models give the relationships between inputs **u**(t) and the states **X**(t), that permit to formulate the direct problem, i.e. the determination of time variation of the states **X**(t) given inputs **u**(t). Control problems require the determination of **u**(t) given a desired time variation of the states **X**(t). This open loop control represents also called inverse problem.

It is easier to illustrate these concepts for a LTI lumped parameters system:

$$d\mathbf{X}(t) / dt = \mathbf{A} \cdot \mathbf{X}(t) + \mathbf{B} \cdot \mathbf{u}(t)$$
$$\mathbf{y}(t) = \mathbf{C} \cdot \mathbf{X}(t)$$

where
X(t) = n-vector of states with given initial conditions **x(0)**
u(t) = m-vector of inputs
y(t) = p-vector of outputs
A, B, C = parameters matrices.

After applying Laplace transform for zero initial conditions this gives

$$(\mathbf{I} \cdot s - \mathbf{A}) \cdot \mathbf{x}(s) = \mathbf{B} \cdot \mathbf{u}(s)$$
$$\mathbf{Y}(s) = \mathbf{C} \cdot \mathbf{x}(s)$$

For this illustration, assuming that all states are directly observable, i.e. **C** = **I**, where **I** is identity matrix, the direct problem is given by

$$\mathbf{x}(s) = (\mathbf{I} \cdot s - \mathbf{A})^{-1} \cdot \mathbf{B} \cdot \mathbf{u}(s)$$

and the solution for inverse problem is

$$\mathbf{u}(s) = \mathbf{B}^{-1} \cdot (\mathbf{I} \cdot s - \mathbf{A}) \cdot \mathbf{x}(s)$$

Assuming as many outputs as states, the inverse problem for desired

$$\mathbf{x}(s) = \mathbf{C}^{-1} \cdot \mathbf{y}_d(s)$$

gives the open loop control law

$$\mathbf{u}(s) = \mathbf{B}^{-1} \cdot (\mathbf{I} \cdot s - \mathbf{A}) \cdot \mathbf{C}^{-1} \cdot \mathbf{y}_d(s)$$

In the path of the solution $\mathbf{u}(s)$ for inverse problem is s and $\mathbf{y}_d(s)$. As a result, fast desired variations of $\mathbf{y}_d(t)$ are subject to a derivative operator and will require extremely high amplitudes of $\mathbf{u}(t)$, leading to an ill-posed problem.

In practice, however, closed loop control is frequently used, with its own advantages and limitations, presented in well known control textbooks [50, 70]. Besides open loop control, inverse problem occurs in numerous other monitoring, identification and estimation problems, investigated in the next chapters of the book.

Direct and inverse problems were extensively investigated since 1960's for various distributed parameters systems, particularly in the case of inverse heat conduction problem [10, 22, 30].

For a generic distributed parameters system, the direct problem, of determining $\mathbf{X}(x, y, z, t)$ given $\mathbf{u}(x, y, z, t)$ and for the noise $\mathbf{w}(x, y, z, t)$, can be formulated as follows

$$\delta\mathbf{X}(x, y, z, t)/\delta t = \mathbf{F}(\mathbf{X}(x, y, z, t), \nabla \mathbf{X}(x, y, z, t), \nabla^2\mathbf{X}(x, y, z, t), \ldots, \mathbf{u}(x, y, z, t), \mathbf{w}(x, y, z, t))$$

$$\mathbf{G}(\mathbf{X}(x, y, z, t), \mathbf{u}(x_b, y_b, z_b, t)) = 0$$

$$\mathbf{I}(\mathbf{X}(x, y, z, 0), \mathbf{u}((x, y, z, 0)) = 0$$

$$\mathbf{H}(\mathbf{y}(x_m, y_m, z_m, t), \mathbf{X}(x, y, z, t)) = 0$$

where

$\mathbf{X}(x, y, z, t)$ is the state vector

$\mathbf{u}(x, y, z, t)$ is the input vector

$\mathbf{H}(\mathbf{y}(x_m, y_m, z_m, t), \mathbf{X}(x, y, z, t)) = 0$ is the output equation defining $\mathbf{y}(x_m, y_m, z_m, t)$, the output vector from measurements from point sensors located at (x_m, y_m, z_m)
$\mathbf{G}(\mathbf{X}(x, y, z, t), \mathbf{u}(x_b, y_b, z_b, t)) = 0$ is the boundary equation for defining the inputs $\mathbf{u}(x_b, y_b, z_b, t)$ from point actuators located at (x_b, y_b, z_b)
$\mathbf{I}(\mathbf{X}(x, y, z, 0), \mathbf{u}((x, y, z, 0)) = 0$ is the initial conditions equation

Control input $\mathbf{u}(x, y, z, t)$ appears either in the boundary conditions $\mathbf{G}(\mathbf{X}(x_b, y_b, z_b, t), \mathbf{u}(x, y, z, t)) = 0$ or in the PDE defining the dynamics $\delta\mathbf{X}/\delta t = \mathbf{F}$. In most cases, control input $\mathbf{u}(x, y, z, t)$ corresponds to point actuators located on the system boundaries. Direct problem of calculating $\mathbf{y}$ and $\mathbf{X}$ given $\mathbf{u}$ is already a difficult to solve problem, analytically or numerically and inverse problem of calculating $\mathbf{u}$ given $\mathbf{y}$ and $\mathbf{X}$ is significantly more difficult to solve. Some simple examples of inverse heat conduction problem will be first used to clarify basic issues.

2.3.2 *Inverse Heat Conduction Problem*

Inverse heat conduction problem is illustrated for the case of a one-dimensional symmetric semi-infinite body, where the heat input q(t) is applied at the boundary x = 0 heated at the boundary x = 0 by the applied heat flux q(t), as shown in Fig. 2.6.

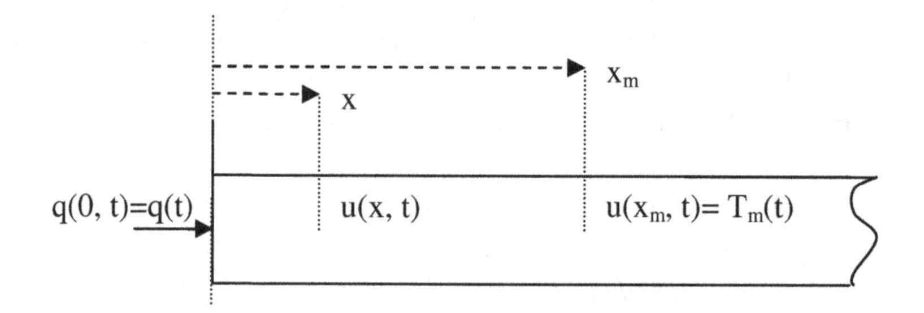

Fig. 2.6 One-dimensional symmetric semi-infinite heated body

This can be modeled, in non-dimensional formulation, by the heat conduction 1D PDE equation for $x > 0$ and $t > 0$ [18, 22]:

$$\frac{\partial^2 u(x,t)}{\partial x^2} = \frac{\partial u(x,t)}{\partial t}$$

or, in a more compact notation

$$u_{xx}(x, t) = u_t(x, t)$$

where $u(x, t)$ is the temperature (in dimensionless units) in point x at time t.

Output equation for exact measurements of the temperature $u(x_m, t)$ from the temperature sensor output $T_m(t)$, located at $x = x_m$, is given by

$$u(x_m, t) = T_m(t)$$

Boundary conditions are given for a semi-infinite body with finite right hand side temperature

$$u(x, t) < \infty \qquad \text{for } x \to \infty$$

and

a) $u(0, t) = T(t)$ in case that the temperature at the left hand side surface ($x = 0$), shown in Fig. 2.6, is the unknown. This is a distant temperature monitoring problem, for the measurement $u(x_m, t) = T_m(t)$.
b) $u_x(0, t) = - q(t)$ in case that the surface heat flux $q(t)$ entering at the left hand side surface of the body ($x = 0$) is the unknown (given that in dimensionless equations the heat flux is proportional to the gradient of the temperature). This is an open loop control problem.

Both a) and b) are inverse problems.
The initial condition is given by

$$u(x, 0) = 0$$

Fourier transform with regard to time gives

$$U_{xx}(x, \omega) = U_t(x, \omega)$$

a) For the case of the above initial and boundary conditions distant temperature monitoring problem, the solution of the above Fourier transform for the Fourier transform of the measured temperature $U(x_m,\omega) = \tau_m(\omega)$ inside the body at sensor location $x = x_m$ given the Fourier transform of the unknown temperature $U(0, \omega)$ at the left hand side surface ($x = 0$) of the body from Fig. 2.6, [22]

$$\tau_m(\omega) = U(0, \omega) \cdot \exp\{-\sqrt{(\omega /2)} \cdot [1 + i \cdot \text{sgn}(\omega)] \}$$

This shows that high frequency components of the surface temperature $U(0, \omega)$ are multiplicd by a term $\exp\{-\sqrt{(\omega /2)} \cdot [1 + i \cdot \text{sgn}(\omega)] \}$ that decreases exponentially with increasing ω, *i.e.* a term that acts as a low pass filter.

The inverse of the above solution gives the Fourier transform of the temperature $U(0, \omega)$ at the left hand side surface ($x = 0$) of the body given the Fourier transform of the measured temperature $\tau_m(\omega)$ inside the body at sensor location $x = x_m$ [22]

$$U(\omega) = \tau_m(\omega) \cdot \exp\{\sqrt{(\omega /2)} \cdot [1 + i \cdot \text{sgn}(\omega)]\}$$

where $\tau_m(\omega)$ is the Fourier transform of the temperature measurements

$$T_m(t) \text{ at } x = x_m$$

This shows that high frequency components of the surface temperature $U(0, \omega)$ are multiplied by a term $\exp\{\sqrt{(\omega /2)} \cdot [1 + i \cdot \text{sgn}(\omega)] \}$ that increases exponentially with increasing ω, *i.e.* a term that acts as high pass filter that reduces the relative weight of useful low frequency components from the measurement signal and amplifies the high frequency components that can contain noise, always present in the output signals of temperature to voltage transducers. This measurement error amplification effect makes the inverse heat conduction problem an ill-posed problem.

b) For this open loop control problem, the boundary condition

$$u_x(0, t) = - q(t)$$

refers the heat flux q(t) entering at the boundary surface (x = 0) that determines the temperature $u(x_m, t)$ inside the body ($x = x_m$), measured as $T_m(t)$. The same type of high pass filter as in case a, leads also to an inverse heat conduction problem that is an ill-posed problem. As a result, open loop control $q^{(c)}(t) = - u_x(0, t)$, based on inverse heat conduction problem formulation, results also in an ill-posed problem.

Same conclusions result from the exact solution of the heat conduction equation

$$\frac{\partial u}{\partial t} = k \cdot \nabla^2 u$$

where u(x, y, z, t) is the temperature in a solid body in the point x, y, z at time t and k is diffusivity.

2.3.3 *Open Loop Control of Distributed Parameters Systems*

Figure 2.7 shows the block diagram for an open-loop control system, when the commands $u^{(c)}(t)$ to the actuators are the result of an open-loop controller.

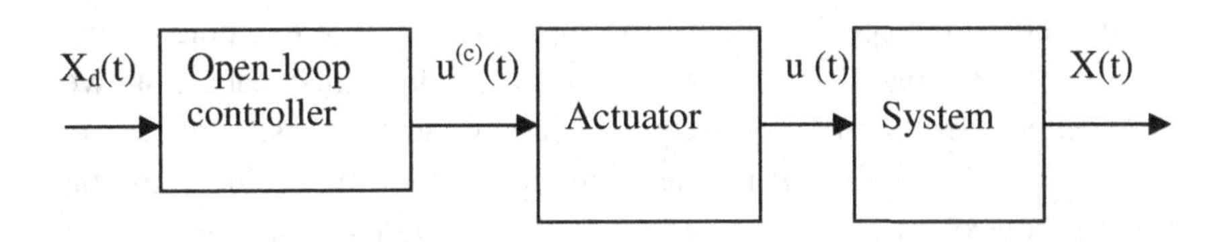

Fig. 2.7 Block diagram for an open-loop system control with no feedback for actuators

In order to reduce the errors between the commands $u^{(c)}(t)$ and outputs of the actuators of the system and actual output of the actuators u(t), an open-loop control system with feedback at actuators is used, as shown in the block diagram from Fig. 2.8.

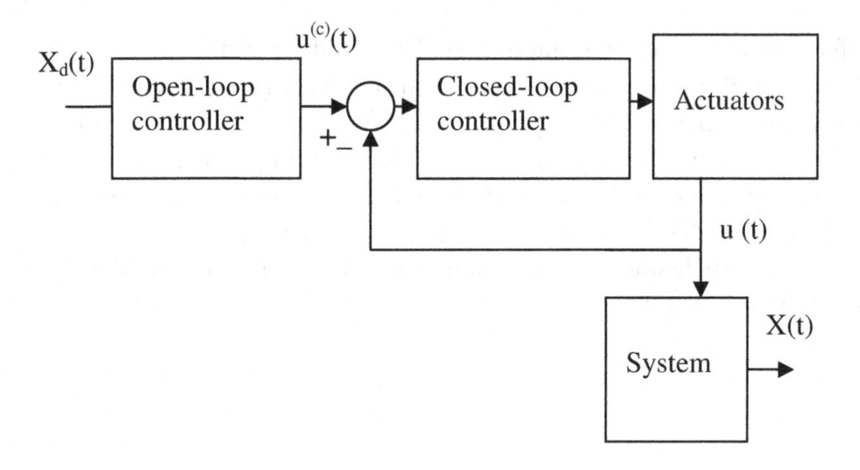

Fig. 2.8 Block diagram for an open-loop control system

In this case a local closed loop control is included for each actuator, much simpler than an overall closed loop control for a distributed parameters system. Distributed parameters systems frequently have coupled dynamics and require a multi input multi output controller, while in the case shown in Fig. 2.8, closed loop control is single input single output for each actuator separately and results in collocated control.

Monitoring and open loop control of distributed parameters systems is affected by the specific inverse problems issues. The example of open loop controller based on the inverse heat conduction equation showed augmented effects of high frequency components, mostly noise, and diminished effects of low frequency components, often more important for the controlled system. In applications, desired internal temperature distributed over space and time is often the result of heat flux from heat sources located at the boundary of the thermal system. The general problem of open loop control of infinite dimensional system with a finite number of point actuators is not yet solved [30]. New results in solving the ill-posed problem accounting for high-frequency components effect, are based on the reformulation of the ill-posed problem into an approximately equivalent well-posed problem [22].

2.3.4 *Closed Loop Control of Distributed Parameters Systems*

Figure 2.9 shows the block diagram for the closed-loop control of a distributed parameters system.

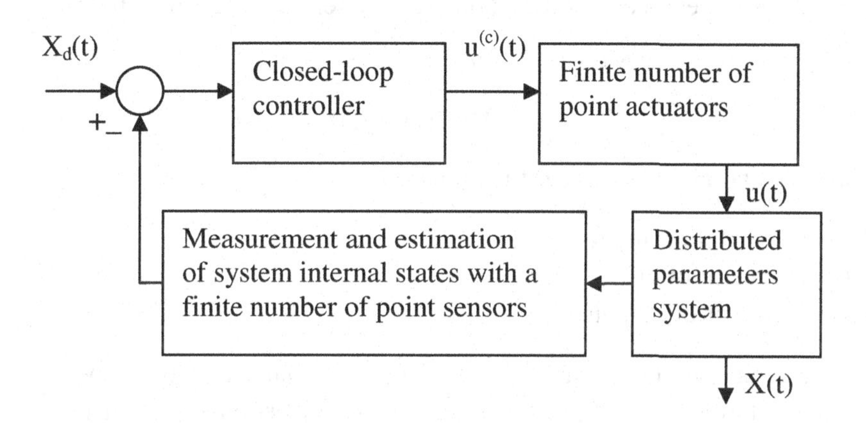

Fig. 2.9 Block diagram for a closed-loop control of a distributed parameters system

This block diagram illustrates the difficulties of the closed-loop control of a distributed parameters system. The measurement and estimation of all system internal states X(x, y, z, t) using a finite number of point sensors, located inside the system or on the boundary, is also an ill-posed problem due to under-sensing and indirect-sensing, as was illustrated above for one-dimensional symmetric semi-infinite heated body. Also, the control of all system internal states X(x, y, z, t) with a finite number of point actuators, located normally on the boundary of the system, is also an infinite dimensional system problem that does not have yet generic satisfactory solutions. Solutions for ill-posed inverse problems are presented in Ch. 3, while their use for particular systems (flexible systems, thermal systems etc) will be presented in the subsequent chapters.

2.4 Under-Actuated and Under-Sensed Mixed Systems

2.4.1 *General Problem of Multi DOF Linear Mechanical Systems. Lumped Parameters Model*

Lumped parameters model for a nonlinear system in the canonic form is:

$$\mathbf{X(t)} / \mathrm{dt} = \mathbf{F}(\mathbf{X(t)}, \mathbf{u}(\mathrm{t}), \mathbf{w}(\mathrm{t}))$$

$$\mathbf{y}(\mathrm{t}) = \mathbf{H}(\mathbf{X(t)}, \mathbf{u}(\mathrm{t}))$$

where **F** and **H** are nonlinear functions and

$\mathbf{u}(t)$ = m-vector of inputs
$\mathbf{w}(t)$ = d-vector of disturbances
$\mathbf{y}(t)$ = p-vector of outputs.

Under-actuation results from fewer inputs m than N, the number of degrees of freedom (DOF), i.e. $m < N$, and under-sensing from fewer outputs p than the number of states n, i.e. $p < n$. Only actuated DOFs can be open loop controlled and only the DOFs with controlled variable measurement by sensors can be closed loop controlled, while the rest of the system will have indirectly controlled dynamics that has to be verified for acceptable bounded states.

Given the complexities of nonlinear models, under-actuation and under-sensing issues are easier to be first analyzed using lumped parameters linear models represented by Ordinary Differential Equations, Linear with Time Invariant parameters (ODE with LTI), presented in Ch. 2.3. Assuming the case that m out of the total of N DOF, ($m < N$), have collocated inputs from actuators and that all these m DOFs have the controlled state variables measured by sensors, than m-N DOF will be un-actuated and the system is both under-actuated and under-sensed. The DOFs subject to closed loop control can have controllers that efficiently bring their state variables towards desired values, while the remaining DOFs can only be subject to redesign to bring their open loop dynamics states within acceptable bounded limits as a result of excitations from other DOFs of the system.

A simple mechanical system will illustrate these issues.

2.4.2 *Two DOF Mechanical System Case*

Figure 2.10 shows a 2 DOF mechanical system.

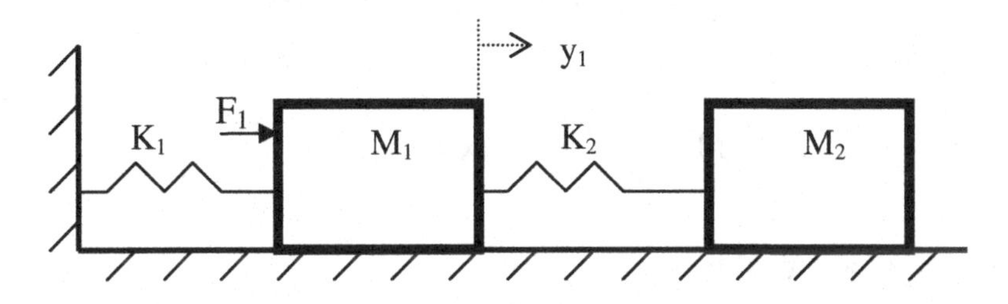

Fig. 2.10 A 2 DOF mechanical system

This 2 DOF mechanical system has one input force F_1 and one position sensor producing the output y_1, and the motion is assumed frictionless. As the result, the degree of freedom corresponding to M_1 is actuated and sensed and the degree of freedom corresponding to M_2 is not actuated and not sensed.

The free body diagrams are shown in Fig. 2.11.

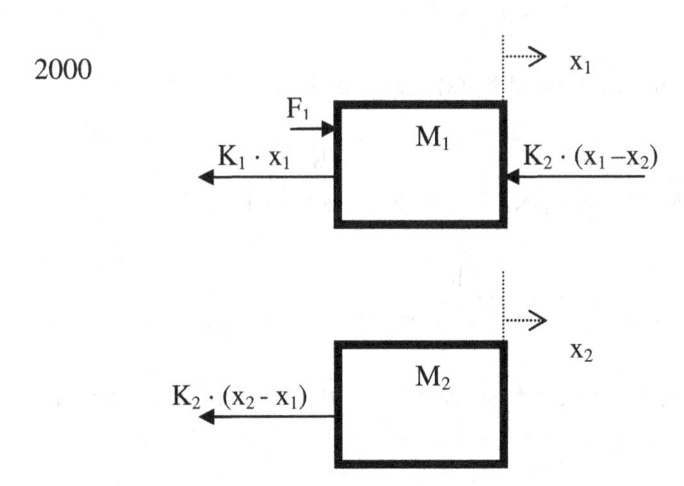

Fig. 2.11 Free body diagrams for the system from Fig. 2.10

The dynamics of this system is described by the following two second order ODEs.

$$M_1 \frac{d^2x_1}{dt^2} + K_1 \cdot x_1 + K_2 \cdot (x_1 - x_2) = F_1$$

$$M_2 \frac{d^2x_2}{dt^2} + K_2 \cdot (x_2 - x_1) = 0$$

The measured output is

$$y_1 = x_1$$

After taking Laplace transform for zero initial conditions, these equations become

$$[M_1 s^2 + K_1 + K_2] \cdot X_1(s) - K_2 \cdot X_2(s) = F_1(s)$$

$$[M_2 s^2 + K_2] \cdot X_2(s) - K_2 \cdot X_1(s) = 0$$

$$Y_1(s) = X_1(s)$$

The position variables $X_1(s)$ and $X_2(s)$ are given by

$$X_1(s) = \frac{1}{M_1 s^2 + K_1 + K_2} F_1(s) + \frac{K_2}{M_1 s^2 + K_1 + K_2} X_2(s)$$

$$X_2(s) = \frac{K_2}{M_2 s^2 + K_2} X_1(s)$$

After replacing $X_2(s)$ from the first equation with the result of the second equation

$$X_1(s) = \frac{1}{M_1 s^2 + K_1 + K_2} F_1(s) + \frac{K_2^{\,2}}{(M_1 s^2 + K_1 + K_2)(M_2 s^2 + K_2)} X_1(s)$$

or

$$X_1(s) = \frac{\dfrac{1}{M_1 s^2 + K_1 + K_2}}{1 - \dfrac{K_2{}^2}{(M_1 s^2 + K_1 + K_2)(M_2 s^2 + K_2)}} F_1(s)$$

The model becomes

$$X_1(s) = \frac{\dfrac{1}{M_1 s^2 + K_1 + K_2}}{1 - \dfrac{K_2{}^2}{(M_1 s^2 + K_1 + K_2)(M_2 s^2 + K_2)}} F_1(s) = G_1(s) \cdot F_1(s)$$

$$X_2(s) = \frac{K_2}{M_2 s^2 + K_2} X_1(s) = G_{12}(s) \cdot X_1(s)$$

$$Y_1(s) = X_1(s)$$

The open loop dynamics model is shown in Fig. 2.12. The consequence is that M_1 is both actuated by F_1 and has the position measured as $y_1 = x_1$. M_1 can have closed loop motion control, while M_2 has open loop dynamics that cannot be directly controlled.

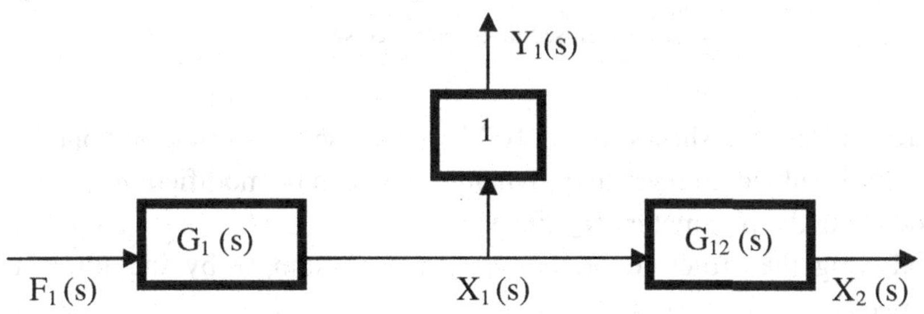

Fig. 2.12 Open loop dynamics diagram

Assume a proportional controller (with gain k_p) of position error of $X_1(s)$ with regard to the desired position $X_{1d}(s)$ giving the force command $F_1^{(c)}(s)$

$$F_1^{(c)}(s) = k_p \cdot (X_{1d}(s) - X_1(s))$$

and an ideal actuator

$$F_1(s) = F_1^{(c)}(s)$$

The closed loop controlled system is shown in Fig. 2.13.

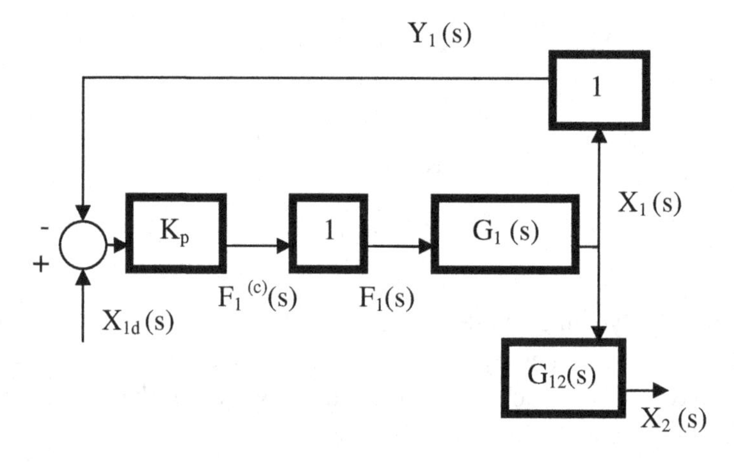

Fig. 2.13 Closed loop controlled system

Figure 2.13 clearly shows that only M_1 has closed loop motion control, while M_2 is subject to open loop dynamics that can be modified only by:

- modifying the parameters K_2 and M_2
- modifying the structure of the system, for example by including a damper B_2
- manipulating the input X_1 (s) to the right hand side open loop system.

In fact, the transfer function for the right hand side open loop system

$$\frac{X_2(s)}{X_1(s)} = \frac{K_2(s)}{M_2 s^2 + K_2} = G_{12}(s)$$

indicates that this system is marginally stable due to the imaginary poles +/- $\sqrt{(K_2/M_2)}$.

This open loop subsystem can be stabilized by modifying the structure of the system, by including a damper B_2, as shown in Fig. 2.14.

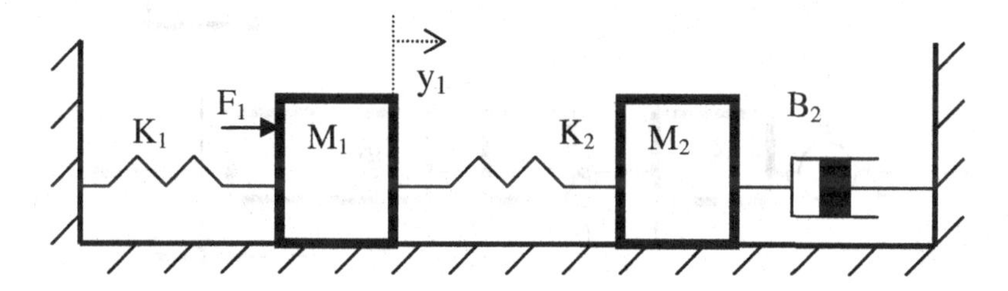

Fig. 2.14 Modified 2 DOF mechanical system

As a result, the equation of motion become

$$M_1 \frac{d^2 x_1}{dt^2} + K_1 \cdot x_1 + K_2 \cdot (x_1 - x_2) = F_1$$

$$M_2 \frac{d^2 x_2}{dt^2} + B_2 \frac{dx_2}{dt} + K_2 \cdot (x_2 - x_1) = 0$$

Following the same procedure as above, the model becomes

$$X_1(s) = \frac{\dfrac{1}{M_1 s^2 + K_1 + K_2}}{1 - \dfrac{K_2{}^2}{(M_1 s^2 + K_1 + K_2)(M_2 s^2 + B_2 s + K_2)}} F_1(s)$$

$$X_1(s) = g_1(s) \cdot F_1(s)$$

$$X_2(s) = \frac{K_2}{M_2 s^2 + B_2 s + K_2} X_1(s) = g_{12}(s) \cdot X_1(s)$$

$$Y_1(s) = X_1(s)$$

The closed loop of the modified controlled system is shown in Fig. 2.15.

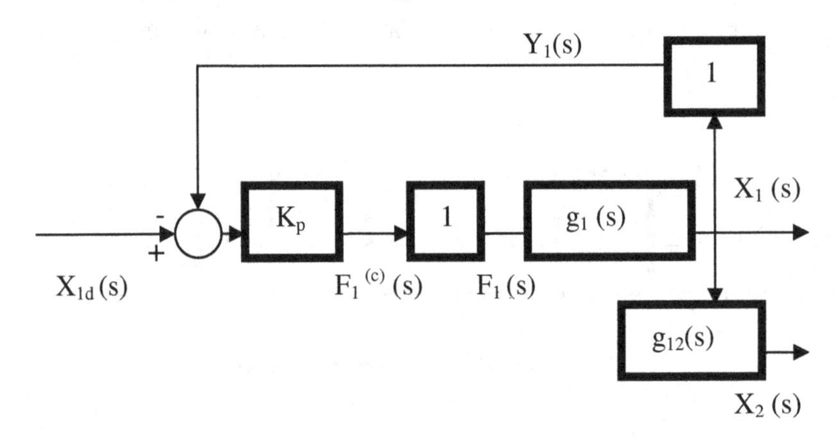

Fig. 2.15 Closed loop control of the modified system

In this case, the transfer function for the right hand side open loop system becomes

$$\frac{X_2(s)}{X_1(s)} = \frac{K_2(s)}{M_2 s^2 + B_2 s + K_2} = g_{12}(s)$$

indicating that this system has poles that can be made under-damped or over-damped by proper choice of the value of B_2 and consequently the open loop subsystem can be maintained within acceptable bounded outputs. Structural system modifications are, however, more expensive than closed loop controllers.

Another case is shown in Fig. 2.16, where M_1 is actuated while only the position of M_2 is sensed. In this case the control of y_2 is indirect.

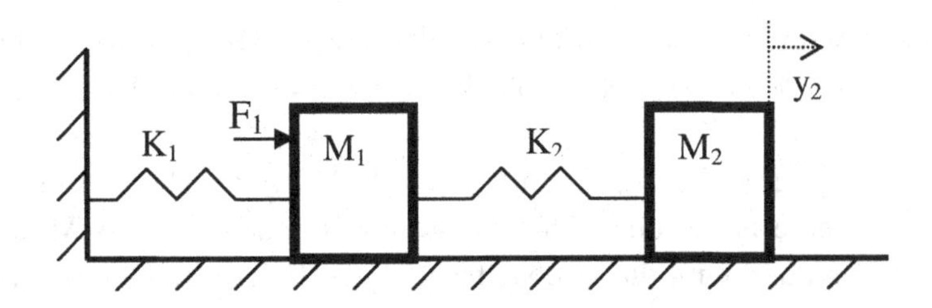

Fig. 2.16 DOF mechanical system with M_1 actuated and M_2 sensed

The closed loop of the system from in Fig. 2.15 is shown in Fig. 2.17.

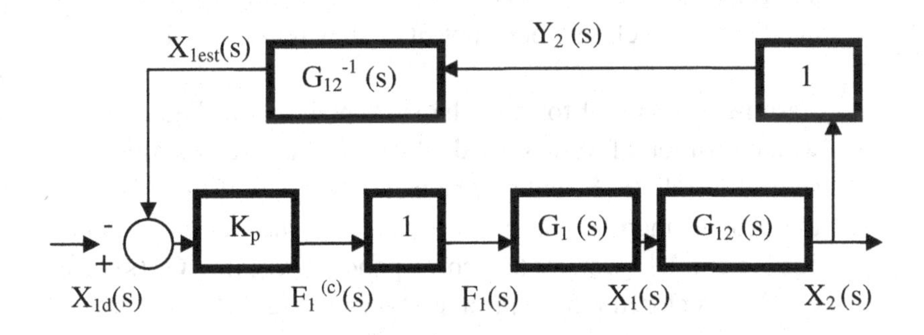

Fig. 2.17 Closed loop control of the system from Fig. 2.16

This control scheme requires the solution $G_{12}^{-1}(s)$ for the inverse problem of estimating $X_{1est}(s)$ given measurements $Y_2(s)$.

Problems

1. For a linear lumped parameters system, which of these is true:
   a) direct and inverse problems are the same
   b) no inverse problem can be defined inverse problem is used for open loop control.

2. Assume that for the 2 DOF system shown in Fig. 2.10 there is viscous friction between the body of mass M_1 and the ground

with the viscous friction coefficient B_1 .Obtain $G_1(s)$ and $G_{12}(s)$ of the corresponding block diagram as shown in Fig. 2.12.

3. Assume that for the modified 2 DOF system shown in Fig. 2.14 there is viscous friction between the body of mass M_1 and the ground with the viscous friction coefficient B_1 .Obtain $g_1(s)$ and $g_{12}(s)$ of the corresponding block diagram closed loop control as shown in Fig. 2.15.

4. Assume a modified form of the system shown in Fig. 2.16, in which position of M_1 is sensed while the mass M_2 is in this case subject to a force F_2. Obtain the corresponding $G_1(s) \cdot G_{12}(s)$ and G_{12}^{-1} for a closed loop control as shown in Fig. 2.17.

5. Assume a modified form of the system shown in Fig. 2.16, in which position of M_1 is sensed while the mass M_2 is in this case subject to a force F_2 and that between the body of mass M_1 and the ground there is viscous friction with the viscous friction coefficient B_1. Obtain the corresponding $G_1(s) \cdot G_{12}(s)$ and G_{12}^{-1} for a closed loop control as shown in Fig. 2.17.

Chapter 3

Overview of Integral Equations and Discrete Inverse Problems

3.1 Integral Equations and Continuous Inverse Problems

3.1.1 *Integral Equations*

Integral equations are considered here for formulating a generic form of models for continuous systems.

Integral equations can be interpreted as forward problems as well as inverse problems, depending on which functions are considered given and which are considered unknown.

Nonlinear Fredholm equation of the first kind [67] is

$$Y(t) = \int_a^b f(t, \tau, U(\tau)) \cdot d\tau$$

This is a definite integral with constant limits a and b. This equation is nonlinear in $U(\tau)$.

If interpreted as a forward problem, $U(\tau)$ is a given continuous function of τ, the input, while $Y(t)$ is the solution of the integration, a continuous function, in this case the unknown output.

If interpreted as an inverse problem, $Y(t)$ is a given continuous function, the input, and $U(\tau)$ is the solution of the integral equation, a continuous function, the unknown input.

Solutions $U(\tau)$ to integral equations represent in this case a generic form of solutions for continuous inverse problems.

Nonlinear Volterra equation of the first kind [67], is a particular form of the nonlinear Fredholm equation, where the limits become $a = -\infty$ and $b = t$. For $\tau > t$, $h(t, \tau) = 0$, which means that $U(\tau)$ has effect on $Y(t)$ only for values prior to τ

$$Y(t) = \int_{-\infty}^{t} f(t, \tau, U(\tau)) \cdot d\tau$$

There is equivalence between Volterra equations and ordinary differential equations, illustrated later as convolution integral.

In order to avoid complexities in solving nonlinear Fredholm equations, this book will focus on Fredholm equation of the first kind linear in $U(\tau)$ [67]

$$Y(t) = \int_{a}^{b} h(t, \tau) \cdot U(\tau) \cdot d\tau$$

For $\tau > t$, $h(t, \tau) = 0$, which means that $U(\tau)$ has effect on $Y(t)$ only for values of t prior to τ.

The kernel $h(t, \tau)$ has often in applications the role of a low pass filter behavior that reduces the effect of fast variations of $U(\tau)$ on $Y(t)$. This is physically due to the effect of inertia in mechanical systems or inductances in electrical systems.

The forward problem solution $Y(t)$ obtained by integration, for a given continuous function $U(\tau)$ is, as a result, a well-posed problem, i.e. it exists, is unique and depends continuously on parameters. In the case of distributed parameters systems, this implies that the corresponding model has suitable initial and boundary conditions [78]. The low pass filter behavior of the kernel $h(t, \tau)$ results also in diminishing the effect of noise content of $U(\tau)$ on the direct problem solution $Y(t)$.

The solution $U(\tau)$ of the inverse problem, for a given continuous function $Y(t)$ is, as a result, a ill-posed problem. If the kernel has a low pass filter behaviour, its effect in the inverse problem is that fast variations of $Y(t)$ due to high frequency noise, result in highly amplified presence in the solution U(τ) and, ultimately in significant difficulties in obtaining numerical solutions [45, 49]. Specialized numerical methods, in particular regularization methods, were developed for this reason.

3.1.2 *Discrete Form*

Integral equations can be approximated by a n-point linear quadrature over the range of integration.

For example, linear Fredholm equation of the first kind

$$Y(t) = \int_a^b h(t, \tau) \cdot U(\tau) \cdot d\tau$$

a n-point linear quadrature has evenly spaced discretization for the range of integration (a, b) with a constant step

$$H = \frac{b - a}{n - 1}$$

For the step count k=0, 1, 2,..., (n-1)

$$t = (K + k) \cdot H$$

$$a = K \cdot H$$

$$b = (K + n - 1) \cdot H$$

Replacing the continuous values for τ from a to b by n-discrete values

$$\tau = K \cdot H, (K+1) \cdot H, \ldots, (K+n-1) \cdot H$$

Fredholm equation can be can be converted using a chosen quadrature in discrete time as follows [45, 49]

$$Y_i = \sum_{k=KH}^{(K+n-1)H} h_{i,k} \cdot u_k$$

After dropping H, Volterra equation can be written in discrete time as follows

$$Y_i = \sum_{k=-\infty}^{i} h_{i,k} \cdot u_k$$

where $h_{i,k}$ are elements of a lower triangular matrix. For discrete Volterra equation u_k is easily computed by forward substitution, given Y_i and $h_{i,k}$ [49].

As a result both Fredholm and Volterra equations can be approximated by systems of linear equations in discrete form leading to the matrix equation

$$\mathbf{Y} = \mathbf{h} \cdot \mathbf{U}$$

where **Y** $[N_y \cdot 1]$, **h** $[N_y \cdot N_u]$ and **U** $[N_u \cdot 1]$.

The coefficients of matrix **h** are obtained for first order or higher order approximations (quadratures) of the continuous function $h(t, \tau)$ for time step H [45, 49].

The calculation is analyzed, for simplicity, for the integral

$$Y = \int_a^b h(\tau) \cdot U(\tau) \cdot d\tau$$

i.e. for the case h dependent only on the integration variable τ that results in a constant value for Y.

The integral is approximated by the sum of a finite number N of terms defined by the discrete values of equally or non-equally spaced the integration variable $\tau_1, \tau_2, ..., \tau_N$

$$Y = \int_a^b h(\tau) \cdot U(\tau) \cdot d\tau = \sum_{j=1}^{N} w(\tau_j) \cdot U(\tau_j)$$

by extending numerical integration rules, for example Simpson rule [45, 49]. The calculation of the weights $w(\tau)$ might be possible analytically or might require again the use of quadrature method.

In the case of Volterra equation in discrete approximation, the matrix **h** is lower triangular and this facilitates significantly the calculation of the solution **h**. Lower triangular structure results in discrete approximation from the continuous form condition $h(t, \tau) = 0$ for $\tau > t$.

In general, the solution of the above matrix equation is obtained using the inverse matrix $\mathbf{h}^{-1}$ for $N_y = N_u$ and non-singular matrix **h**

$$\mathbf{U} = \mathbf{h}^{-1} \cdot \mathbf{Y}$$

In monitoring, estimation, identification and control applications, matrix **h** $[N_y \cdot N_u]$ is rarely square or non-singular, and the solution of the matrix equation uses the generalized inverse $\mathbf{h}^{-g}$ $[N_u \cdot N_y]$

$$\mathbf{U} = \mathbf{h}^{-g} \cdot \mathbf{Y}$$

3.1.3 *Other Examples of Discrete Inverse Problems*

Discrete inverse problems do not result only from discretization of integral equations. In this section curve fitting case will be investigated. Boundary measurement of single ray reflection in distributed parameters systems and boundary measurement of direct ray propagation in distributed parameters systems are presented in detail in specialized texts [45, 65, 67]. Linear and nonlinear curve fitting try to recover a continuous functional dependence of variables given a set of discrete data.

Example 3.1 Assume that for the experimental data for the measured resistance versus temperature consists of $m = 6$ pairs of values

i	1	2	3	4	5	6
R_i [Ω]	600	627	651	682	701	732
T_i [^{0}C]	24	25.2	27	28.7	29	29.8

For linear curve fitting the following algebraic equation, an assumed direct problem model is considered

$$R = a_0 + a_1 \cdot T$$

with unknown parameters a and b. This leads to an over-determined inverse problem for n = 2 unknowns and m = 6 equations. The over-determined matrix equation is in this case

$$\mathbf{Y} = \mathbf{h} \cdot \mathbf{U}$$

where the vector of unknowns is

$$\mathbf{U} = \begin{bmatrix} a_0 \\ a_1 \end{bmatrix}$$

For **Y** [6 · 1], **h** [6 · 2] and **U** [2 · 1], the matrix equation becomes

$$\begin{bmatrix} 600 \\ 627 \\ 651 \\ 682 \\ 701 \\ 732 \end{bmatrix} = \begin{bmatrix} 1 & 24.0 \\ 1 & 25.2 \\ 1 & 27.0 \\ 1 & 28.7 \\ 1 & 29.0 \\ 1 & 29.8 \end{bmatrix} \cdot \begin{bmatrix} a_0 \\ a_1 \end{bmatrix}$$

Inverse problem in this case consists in determining the unknown parameters a_0 and a_1 of the assumed inherent linear dependence between R and T of a direct problem.

Solving the linear regression problem consists in obtaining the unknown parameters by minimizing the Euclidean norm of the errors, i.e. in calculating the Least Squares (LS) error solution [28]

$$\min S(a_0, a_1) = \min \sum_{i-1}^{m} \{R_i - a_1 T_i - a_0\}^2$$

The solution is obtained from vanishing partial derivatives

$$\frac{\partial S}{\partial a_1} = \sum_{i-1}^{m} 2 \cdot \{R_i - a_1 T_i - a_0\} \cdot (-T_i) = 0$$

$$\frac{\partial S}{\partial a_0} = \sum_{i-1}^{m} 2 \cdot \{R_i - a_1 T_i - a_0\} \cdot (-1) = 0$$

or

$$a_1 \sum_{i-1}^{m} T_i^2 + a_0 \sum_{i-1}^{m} T_i = \sum_{i-1}^{m} T_i R_i$$

$$a \sum_{i-1}^{m} T_i + b \cdot m = \sum_{i-1}^{m} R_i$$

The following matrix equation results for n=2 unknowns, a and b,

$$\begin{bmatrix} m & \sum_{i-1}^{m} T_i \\ \sum_{i-1}^{m} T_i & \sum_{i-1}^{m} T_i^2 \end{bmatrix} \begin{bmatrix} a_0 \\ a_1 \end{bmatrix} = \begin{bmatrix} \sum_{i-1}^{m} R_i \\ \sum_{i-1}^{m} T_i R_i \end{bmatrix}$$

Inverse problem solution is given by

$$\begin{bmatrix} a_0 \\ a_1 \end{bmatrix} = \begin{bmatrix} m & \sum_{i-1}^{m} T_i \\ \sum_{i-1}^{m} T_i & \sum_{i-1}^{m} T_i^2 \end{bmatrix}^{-1} \begin{bmatrix} \sum_{i-1}^{m} T_i R_i \\ \sum_{i-1}^{m} R_i \end{bmatrix}$$

For nonlinear curve fitting the following n-order polynomial can be chosen

$$R = a_0 \cdot T^0 + a_1 \cdot T^1 + \ldots + a_n \cdot T^n$$

The inverse problem consists in this case in calculating the unknown n-parameters, $a_0, a_1, \ldots, a_n$ given data pairs

$$(R_1, T_1), (R_2, T_2), \ldots, (R_m, T_m)$$

The inverse problem is normally over-determined, m > n, *i.e.* there are more data pairs m than unknowns n. The solution can be obtained again using a least squares error solution. The above linear curve fitting problems will be solved in Ch. 3.3 using a MATLAB function.

Other examples of discrete inverse problems, boundary measurement of single ray reflection in distributed parameters systems, boundary measurement of direct ray propagation in distributed parameters systems *etc.* will be presented in next chapters of the book.

3.2 Discrete Problems for LTI Systems

3.2.1 *Introduction*

Inverse problems for Linear Time Invariant (LTI) systems can be formulated for different representations of the forward model of the system, for example for lumped parameters systems or for distributed parameters systems.

In what follows the goal is to present solutions to inverse problems that are numerically efficient and common to both for lumped parameters systems and distributed parameters systems. In subsequent chapters, a particular attention will be given to problems requiring real-time solution, for example for dynamic compensation of sensors.

3.2.2 *Lumped Parameters Systems*

Inverse problems for LTI lumped parameters systems, for the case of fewer inputs and fewer measurements than states (more representative for lumped parameters approximation of a distributed parameters system), for different representations of the forward model of the system can be formulated as follows:

3.2.2.1 *State space representation*

State dynamics and output LTI ODE equations from Ch. 2.3 are completed here with measurement noise w(t):

$$d\mathbf{X}(t) / dt = \mathbf{A} \cdot \mathbf{X}(t) + \mathbf{B} \cdot \mathbf{u}(t)$$
$$\mathbf{y}_m(t) = \mathbf{C} \cdot \mathbf{X}(t) + \mathbf{w}(t)$$

where
$\mathbf{X}(t)$ = n-vector of states with given initial conditions x(0)
$\mathbf{u}(t)$ = m-vector of inputs
$\mathbf{y}(t)$ = p-vector of measurement outputs
$\mathbf{w}(t)$ =p-vector of measurement noise
$\mathbf{A}, \mathbf{B}, \mathbf{C}$ = parameters matrices.
and $m < n$ and $p < n$

We denote the noiseless output as

$$\mathbf{y}(t) = \mathbf{y}_m(t) - \mathbf{w}(t)$$

such that the output equation becomes

$$\mathbf{y}(t) = \mathbf{C} \cdot \mathbf{X}(t)$$

In this case the inverse problem of estimating the input, $\mathbf{u}_{est}(t)$, from measurements $\mathbf{y}(t)$ is obtained by solving first the output equation for $m < n$ using the generalized inverse $\mathbf{C}^{-g}$

$$\mathbf{X}(t) = \mathbf{C}^{-g} \cdot \mathbf{y}(t)$$

and then

$$d\mathbf{X}(t)/dt = \mathbf{C}^{-g} \cdot d\mathbf{y}(t) / dt$$

that gives

$$\mathbf{C}^{-g} \cdot d\mathbf{y}(t)/dt = \mathbf{A} \cdot \mathbf{C}^{-g} \cdot \mathbf{y}(t) + \mathbf{B} \cdot \mathbf{u}(t)$$

The solution for u(t) of the state equation gives

$$\mathbf{u}_{est}(t) = \mathbf{B}^{-g} \cdot (d\mathbf{X}(t) / dt - \mathbf{A} \cdot \mathbf{X}(t))$$

or

$$\mathbf{u}_{est}(t) = \mathbf{B}^{-g} \cdot (\mathbf{C}^{-g} \cdot d\mathbf{y}(t) / dt - \mathbf{A} \cdot \mathbf{C}^{-g} \cdot \mathbf{y}(t))$$

This solution requires the calculation of generalized inverses $\mathbf{b}^{-g}$ and $\mathbf{c}^{-g}$ as well as the derivative $d\mathbf{y}(t)/dt$. Real-time implementation of this solution is often computationally intensive and requires specific code for each application. The presence of noise and discontinuities in $\mathbf{y}_{\mathbf{m}}(t)$ signal might lead to unacceptable signal to noise ratios, which is not suitable for real-time implementation.

3.2.2.2 *Complex functions representation*

After applying Laplace transform to the above LTI ODE system for zero initial conditions the result is

$$(\mathbf{I} \cdot s - \mathbf{A}) \cdot \mathbf{x}(s) = \mathbf{B} \cdot \mathbf{u}(s)$$
$$\mathbf{y}(s) = \mathbf{C} \cdot \mathbf{x}(s)$$

where

$$\mathbf{y}(s) = \mathbf{y_m}(s) - \mathbf{w}(s)$$

Assuming that all states are directly observable, i.e. $\mathbf{C} = \mathbf{I}$, where $\mathbf{I}$ is identity matrix, the direct problem is given by

$$\mathbf{x}(s) = (\mathbf{I} \cdot s - \mathbf{A})^{-1} \mathbf{B} \cdot \mathbf{u}(s)$$

while the solution for the inverse problem is

$$\mathbf{u}(s) = \mathbf{B}^{-1} \cdot (\mathbf{I} \cdot s - \mathbf{A}) \cdot \mathbf{x}(s)$$

Assuming as many outputs as states, the inverse problem solution for desired output $\mathbf{y}_d$ is

$$\mathbf{x}\,(s) = \mathbf{C}^{-1} \cdot \mathbf{y}_d\,(s)$$

and the open loop control law is

$$\mathbf{u}_{est}\,(s) = \mathbf{B}^{-1} \cdot (\mathbf{I} \cdot s - \mathbf{A}) \cdot \mathbf{C}^{-1} \cdot \mathbf{y}_d\,(s)$$

This solution requires also the calculation of generalized inverses $\mathbf{B}^{-g}$ and $\mathbf{C}^{-g}$. The presence of "s" in the feed-forward path from $\mathbf{y}(s)$ to $\mathbf{u}(s)$, corresponds to time derivative of $\mathbf{y}_d$. As a result, real-time implementation of this solution is not desirable.

3.2.2.3 *Convolution integral representation*

Convolution integral representation is of interest as a link to non-linear forward problems formulation using integral equations and as a basis for developing computationally efficient matrix formulation.

The principle of superposition, valid for linear systems, gives [70]

$$Y(t) = \int_{-\infty}^{t} h(t,\tau) \cdot U(\tau) \cdot d\tau$$

where h(t, τ) is the impulse response of the system, for the impulse assumed applied at any time $\tau \leq t$. This is a Volterra equation of the first kind and $h(t, \tau) = 0$ for $\tau > t$ [9].

In the case of LTI systems,

$$h(t, \tau) = h(t - \tau)$$

and depends only on the difference between the time τ, when the impulse is applied, and the time t, when the response y is observed. This property greatly reduces the computation of the impulse response h. The convolution integral for LTI systems is given by

$$Y(t) = \int_{-\infty}^{t} h(t-\tau) \cdot U(\tau) \cdot d\tau$$

The calculation of the impulse response for a system modeled by a LTI ODE model results from considering a unit impulse input $\mathbf{U}(t) = \boldsymbol{\delta}(t)$, such that

$$d\mathbf{X}(t) / dt = \mathbf{A} \cdot \mathbf{X}(t) + \mathbf{B} \cdot \boldsymbol{\delta}(t)$$

and

$$\mathbf{C}^{-g} \cdot d\mathbf{y}(t) / dt = \mathbf{A} \cdot \mathbf{C}^{-g} \cdot \mathbf{y}(t) + \mathbf{B} \cdot \boldsymbol{\delta}(t)$$

Rather then solving analytically this equation for $\mathbf{y}(t) = \mathbf{h}(t)$, complex functions representation can be used to obtain the transfer function. For $n = m = p = 1$

$$h(s) = \frac{y(s)}{u(s)}$$

and for the unit impulse input u(s) =1, impulse response h(s) is

$$h(s) = y(s)$$

In time domain, inverse Laplace transform L^{-1} gives

$$h(t) = L^{-1}\ \{h(s)\} \quad \text{for } t > 0$$

The condition $t > 0$ reflects the property that the input signals starting at $t = 0$ cannot affect the output for $t < 0$.

Examples of impulse responses are [70]:

A) First order system with the transfer function

$$h(s) = \frac{1}{s+a}$$

which has the inverse Laplace transform

$$h(t) = e^{-at}\ 1(t)$$

where 1(t) is unit step function indicating that an input applied at $t = 0$ cannot influence the output for $t < 0$.

B) Second order system with the transfer function

$$h(s) = \omega_n^{\ 2} / \left(s^2 + 2 \cdot \zeta \cdot \omega_n \cdot s + \omega_n^{\ 2} \right)$$

has the inverse Laplace transform

$$h(t) = \frac{\omega_n}{\sqrt{1-\zeta^2}} [e^{-\zeta\omega_n t} \cdot \sin(\sqrt{1-\zeta^2} \cdot \omega_n \cdot t)] \cdot 1(t)$$

C) Higher order systems transfer functions can be expanded by partial fraction expansion to a sum of first and second order transfer functions for which impulse responses can be determined by inverse Laplace transform as above.

Convolution integral can be reformulated in the discrete form convolution sum using shifted impulse response h_{i-j}, corresponding to the sampled time interval t - τ with sampling period T_s. The discrete time t_i, (corresponding to the continuous time τ) when the impulse is applied [67], is given by

$$t_i = i \cdot T_s$$

and the time t_i when the response y is observed

$$t_j = j \cdot T_s$$

such that t - τ in discrete time is (i - j) T_s, or i - j in steps.

Volterra equation of the first kind [67]

$$Y(t) = \int_{-\infty}^{t} h(t,\tau) \cdot U(\tau) \cdot d\tau$$

can be written in discrete time as follows

$$Y_i = \sum_{k=-\infty}^{i} h_{i,k} \cdot u_k$$

where $h_{i,k}$ are terms of a lower triangular matrix. In this case Y_i is easily computed by forward substitution given $h_{i,k}$ and u_k [49].

Convolution sum for LTI discrete systems is

$$Y(t) = \int_{-\infty}^{t} h(t-\tau) \cdot U(\tau) \cdot d\tau$$

which can be written in discrete time as follows

$$Y_i = \sum_{k=-\infty}^{i} h_{i-k} \cdot u_k$$

where the condition embodied in continuous time in 1(t) is replaced here by $i - k \geq 0$ or $k \leq i$.

For LTI systems, a recursive formula can be obtained using the property $h_{i-k} = h_{i-1-k}\, h_1$. For the current time i, the output depends on the effect of the current input u_i and also on all previous inputs u_k (for $k = -\infty$ to $i - 1$) weighted by the corresponding discrete impulse response h_{i-k} values

$$Y_i = \sum_{k=-\infty}^{i} h_{i-k} \cdot u_k = \sum_{k=-\infty}^{i-1} h_{i-k} \cdot u_k + h_o \cdot u_i$$

or, the recursive formula

$$Y_i = Y_{i-1} + h_0 \cdot u_i$$

where

$$Y_{i-1} = \sum_{k=-\infty}^{i-1} h_{i-k} \cdot u_k$$

The impulse response in discrete time t_i, for first order systems is

$$h_i = e^{-at_i} \qquad \text{for } i \geq 0$$

and for second order systems is

$$h_i = \frac{\omega_n}{\sqrt{1-\zeta^2}} \cdot [e^{-\zeta\omega_n t_i} \cdot \sin(\sqrt{1-\zeta^2} \cdot \omega_n \cdot t_t)] \qquad \text{for } i \geq 0$$

For actual physical systems, input signals are of limited duration and the significant part of the transient response is characterized by the finite settling time. Also, the output reflects both the effect of the periodic input with period $T_i = 2 \cdot \pi / \omega_i$ and of the under-damped response of the system with natural period $T_n = 2 \cdot \pi / \omega_n$. Assume that an earlier time $t_{i-K} = (i-K) \cdot T_s$ of an older input can still affect current time t_i output. For current time step i = 0, i.e. at time t_0, the oldest input signal still affecting current output can be assumed to having been applied at –K time steps, i.e. at $t_{-K} = -K \cdot T_s$. As a result, at the current time step i, the summation can be limited to $i - K < k \leq i$, *i.e.*

$$Y_i = \sum_{k=i-K}^{i} h_{i-k} \cdot u_k \quad \text{for } i \geq 0$$

At the lower limit k = i - K, the input u_{i-K} is multiplied h_K to determine its contribution to the current output Y_i. For i = 0, the product is $u_{-K} \cdot h_K$ and determines the contribution of u_{-K} to the output Y_0. In this case, inverse problem solving is reduced a finite number of K forward substitutions [49].

This result corresponds to the calculation of the sum for a sliding window of K steps from the past to the current time.

In this book, the input $u(\tau)$ is assumed applied only at $\tau \geq 0$, similar to the condition for unilateral Laplace transform. In discrete time this means that $u_k = 0$ for $k < 0$ and Y_i depends only on the input values u_k at k = 0, 1, 2,….,i.

Example 3.2 Assume 2% criterion for first order system $h(t) = e^{-at}\, 1(t)$ for the duration of the settling time. For a = 0.2, the time constant is $\tau = 1 / a = 5$ [s], and sampling time is chosen less than half the time constant, in this case $T_s = 1$. Over $4 \cdot \tau$ for 2% criterion, *i.e.* for 20 [s], at sampling time $T_s = 1$, there are i = 1, 2,…, 20 discrete time steps.

Impulse response for the first order system

$$h_i = e^{-at_i} \text{ for } i \geq 0$$

gives

$$h_0 = \exp\{-0.2 \cdot 0\} = 1$$
$$h_1 = \exp\{-0.2\} = 0.82$$
$$h_2 = \exp\{-0.4\} = 0.67$$
$$h_3 = \exp\{-0.6\} = 0.55$$
$$h_4 = \exp\{-0.8\} = 0.45$$
$$h_5 = \exp\{-1.0\} = 0.37$$
$$\cdots\cdots\cdots\cdots$$
$$h_{10} = \exp\{-2\} = 0.13$$
$$------------$$
$$h_{20} = \exp\{-4\} = 0.018$$

Consequently, after 20 time steps the impulse response is less than 2% from the steady state value and the rest of the transient regime can be ignored in practical applications.

Example 3.3 For the previous system, for $i \geq k$ or $i - k \geq 0$ and $i \geq 0$

$$Y_i = \sum_{k=-i-K}^{i} h_{i-k} \cdot u_k$$

has nonzero products values only for $i \geq k \geq 0$. Assume the input

$$u_k = 0 \text{ for } k < 0$$
$$u_k = 10 \text{ for } k = 0$$
$$u_k = 0 \text{ for } k > 0$$

such that the only non-zero value is $u_0 = 10$.

The first 20 time steps outputs are:

$$Y_0 = h_0 \cdot u_0 = 10$$
$$Y_1 = h_1 \cdot u_0 = \exp\{-0.2\} \cdot 10 = 8.2$$
$$\cdots\cdots\cdots\cdots$$

$$Y_{10} = h_{10} \cdot u_0 = \exp\{-2\} \cdot 10 = 1.3$$

$$Y_{20} = h_{20} \cdot u_0 = \exp\{-4\} \cdot 10 = 0.18$$

These results illustrate that the the single non-zero input value, $u_0 = 10$ has diminishing in time effect on Y_i for i = 1.2,..., 20.

Example 3.4 For the previous system, assume the input

$$u_k = 0 \text{ for } k < 0$$
$$u_0 = 10$$
$$u_1 = 1$$
$$u_k = 0 \text{ for } k > 1$$

The output is

$$Y_i = \sum_{k=i-K}^{i} h_{i-k} \cdot u_k = h_i \cdot u_0 + h_{i-1} \cdot u_1$$

is

$$Y_0 = h_0 \cdot u_0 = 10$$
$$Y_1 = h_1 \cdot u_0 + h_0 \cdot u_1 = \exp\{-0.2\} \cdot 10 + 1 = 9.65$$

....

$$Y_{10} = h_{10} \cdot u_0 + h_9 \cdot u_1 = \exp\{-2\} \cdot 10 + \exp\{-1.8\} = 1.75$$

$$Y_{20} = h_{20} \cdot u_0 + h_{19} \cdot u_1 = \exp\{-4\} \cdot 10 + \exp\{-3.8\} = 0.2$$

The result is the superposition of the outputs for u_0 and u_1.

3.2.2.4 *Matrix form representation*

Matrix form representation of discrete Volterra equation forward model is

$$\mathbf{Y} = \mathbf{h} \cdot \mathbf{U}$$

where
U is input $[N_u \cdot 1]$ vector
Y(t) is noiseless output $[N_y \cdot 1]$ vector
h is $[N_u \cdot N_y]$ matrix

and

$$h_{i,k} = h_{i-k} \text{ for } i - k \geq 0$$

and

$$h_{i,k} = 0 \text{ for } i - k < 0$$

such that

$$h_{i,k} = 0 \quad \text{for } i < k$$
$$h_{i,k} = h_0 \quad \text{for } i = k$$
$$h_{i,k} = h_1 \quad \text{for } i = k + 1$$
$$\cdots\cdots\cdots\cdots\cdots$$
$$h_{i,k} = h_j \text{ for } i = k + j$$

etc.

Matrix representation permits to define easily the inverse problem of determining the values of the input u_k given the values of the output Y_i as long as the matrix h is known.

For a non-invertable matrix **h**,

$$\mathbf{U} = \mathbf{h}^{-g} \cdot \mathbf{Y}$$

Three cases can be identified:
A) $N_y = N_u$,
B) $N_y < N_u$
C) $N_y > N_u$

$N_y = N_u = K+1$

For this case, the matrix **h** [(K+1) · (K+1)] is a lower triangular square matrix

$$\mathbf{h} = \begin{bmatrix} h_o & 0 & 0 & . & 0 & . & 0 & 0 \\ h_1 & h_o & 0 & . & 0 & . & 0 & 0 \\ h_2 & h_1 & h_o & . & 0 & . & 0 & 0 \\ . & . & . & . & . & . & . & . \\ h_j & h_{j-1} & h_{j-2} & . & h_o & . & 0 & 0 \\ . & . & . & . & . & . & . & . \\ . & . & . & . & . & . & . & . \\ h_K & h_{K-1} & h_{K-2} & . & . & . & h_1 & h_o \end{bmatrix}$$

The solution **U** of equation $\mathbf{Y} = \mathbf{h} \cdot \mathbf{U}$ for given **Y** and **h** [(K +1) · (K +1)] will have the same length K+1 as the given **Y** and can be calculated continuously, as a length K +1 sliding window ending at current time i and for the input u_k for $i\text{-}K \leq k \leq i$. The input might have, however, several zero values. Moreover, for the case of a lower triangular matrix h, for the direct problem

$$\mathbf{Y} = \mathbf{h} \cdot \mathbf{U}$$

or for

$$Y_i = \sum_{k=i-K}^{i} h_{i-k} \cdot u_k$$

given **Y** and **h**, the solution **U** of is easily computed by forward substitution [49]. This permits to solve analytically the inverse problem. For $u_k = 0$ for $k < 0$, scalar forward model is

$$Y_0 = h_0 \cdot u_0$$

$$Y_1 = h_1 \cdot u_0 + h_0 \cdot u_1$$
$$Y_2 = h_2 \cdot u_0 + h_1 \cdot u_1 + h_0 \cdot u_2$$
$$\dots\dots\dots\dots$$
$$Y_i = h_i \cdot u_0 + h_{i-1} \cdot u_1 + h_{i-2} \cdot u_2 + \dots\dots h_0 \cdot u_i$$
$$\dots\dots\dots\dots$$
$$Y_K = h_K \cdot u_0 + h_{K-1} \cdot u_1 + \dots \dots\dots h_0 \cdot u_K$$

or, in matrix form, for the a lower triangular matrix **h**

$$\begin{bmatrix} Y_0 \\ Y_1 \\ Y_2 \\ \cdot \\ Y_j \\ \cdot \\ \cdot \\ Y_K \end{bmatrix} = \begin{bmatrix} h_o & 0 & 0 & . & 0 & . & 0 & 0 \\ h_1 & h_o & 0 & . & 0 & . & 0 & 0 \\ h_2 & h_1 & h_o & . & 0 & . & 0 & 0 \\ \cdot & \cdot & \cdot & \cdot & \cdot & \cdot & \cdot & \cdot \\ h_j & h_{j-1} & h_{j-2} & . & h_o & . & 0 & 0 \\ \cdot & \cdot & \cdot & \cdot & \cdot & \cdot & \cdot & \cdot \\ \cdot & \cdot & \cdot & \cdot & \cdot & \cdot & \cdot & \cdot \\ h_K & h_{K-1} & h_{K-2} & \cdot & \cdot & \cdot & h_1 & h_o \end{bmatrix} \cdot \begin{bmatrix} u_0 \\ u_1 \\ u_2 \\ \cdot \\ u_i \\ \cdot \\ \cdot \\ u_K \end{bmatrix}$$

The direct problem result $\mathbf{Y} = \mathbf{h} \cdot \mathbf{U}$ is easy to obtain given **h** and **U**. The inverse problem solution **U** given **h** and **Y**, is normally more difficult to obtain. The calculation of **U** can be carried out using the inverse problem formulation for a non-singular square matrix **h**

$$\mathbf{U} = \mathbf{h}^{-1} \cdot \mathbf{Y}$$

For LTI systems, given that **h** is constant, numerical inversion of **h** into $\mathbf{h}^{-1}$ of can be done off-line. In this case the computation of K+1 values of Y_i for i = 0, 1,..., K consists in a weighted sum of current values of u_i, i = 1, 2,..., (K+1) with known constant values of the elements of $\mathbf{h}^{-1}$ [(K+1)·(K+1)]. This shows that after the calculation of $\mathbf{h}^{-1}$, obtaining the result for **U** is a simple summation of weighted inputs.

Example 3.5 Assume a discrete first order system with K = 3 and a = 0.2 [s^{-1}] with

$$h_i = e^{-at_i} \text{ for } i=0,1,...3$$

Obtain

a) the non-zero elements of the discrete impulse response matrix
h [4 · 4] and **Y** = **h** · **U**

b) Obtain **Y** for

$$\begin{bmatrix} u_0 \\ u_1 \\ u_2 \\ u_3 \end{bmatrix} = \begin{bmatrix} 0 \\ 10 \\ 0 \\ 0 \end{bmatrix}$$

c) Obtain **Y** for

$$\begin{bmatrix} u_0 \\ u_1 \\ u_2 \\ u_3 \end{bmatrix} = \begin{bmatrix} 0 \\ 10 \\ 1 \\ 0 \end{bmatrix}$$

a) The non-zero elements of the discrete impulse response are

$$h_0 = \exp\{-0.2 \cdot 0\}=1$$
$$h_1 = \exp\{-0.2\}=0.82$$
$$h_2 = \exp\{-0.4\}=0.67$$
$$h_3 = \exp\{-0.6\}=0.55$$

The [4 · 4] matrix equation

$$\begin{bmatrix} Y_0 \\ Y_1 \\ Y_2 \\ Y_3 \end{bmatrix} = \begin{bmatrix} h_0 & 0 & 0 & 0 \\ h_1 & h_0 & 0 & 0 \\ h_2 & h_1 & h_0 & 0 \\ h_3 & h_2 & h_1 & h_0 \end{bmatrix} \cdot \begin{bmatrix} u_0 \\ u_1 \\ u_2 \\ u_3 \end{bmatrix}$$

in this case is

$$\begin{bmatrix} Y_0 \\ Y_1 \\ Y_2 \\ Y_3 \end{bmatrix} = \begin{bmatrix} 1 & 0 & 0 & 0 \\ 0.82 & 1 & 0 & 0 \\ 0.67 & 0.82 & 1 & 0 \\ 0.55 & 0.67 & 0.82 & 1 \end{bmatrix} \cdot \begin{bmatrix} u_0 \\ u_1 \\ u_2 \\ u_3 \end{bmatrix}$$

b) For

$$\begin{bmatrix} u_0 \\ u_1 \\ u_2 \\ u_3 \end{bmatrix} = \begin{bmatrix} 0 \\ 10 \\ 0 \\ 0 \end{bmatrix}$$

the direct problem is

$$\begin{bmatrix} Y_0 \\ Y_1 \\ Y_2 \\ Y_3 \end{bmatrix} = \begin{bmatrix} 1 & 0 & 0 & 0 \\ 0.82 & 1 & 0 & 0 \\ 0.67 & 0.82 & 1 & 0 \\ 0.55 & 0.67 & 0.82 & 1 \end{bmatrix} \cdot \begin{bmatrix} 0 \\ 10 \\ 0 \\ 0 \end{bmatrix}$$

and the result Y of the direct problem Y =h · U given U is

$$Y_0 = 0$$
$$Y_1 = 10$$
$$Y_2 = 8.2$$
$$Y_3 = 6.7$$

or

$$\begin{bmatrix} Y_0 \\ Y_1 \\ Y_2 \\ Y_3 \end{bmatrix} = \begin{bmatrix} 0 \\ 10 \\ 8.2 \\ 6.7 \end{bmatrix}$$

As expected, for a first order system subject to a delayed impulse input by one time step, an exponentially decaying output, also delayed by one time step, is obtained. Due to the limitation of the output to only K = 3, a significant part of the output, beyond Y_3, is not observed.

Normally, for this system with a time constant $\tau = 1/a = 5$ [s], 2% settling time requires output observations for 4 time constants, *i.e.* 20 [s]. The time constant of 5 [s] requires a sampling time significantly shorter, for example of 1 [s]. This would require measurements of the output for K = 20 time steps, much longer than the above K = 3.

In the case of an input **U** [20 · 1], instead of the above [4 · 1], output values for **Y** [20 · 1] would be obtained with $\mathbf{Y} = \mathbf{h} \cdot \mathbf{U}$.

c) In this case, for

$$\begin{bmatrix} u_0 \\ u_1 \\ u_2 \\ u_3 \end{bmatrix} = \begin{bmatrix} 0 \\ 10 \\ 1 \\ 0 \end{bmatrix}$$

the direct problem is given by

$$\begin{bmatrix} Y_0 \\ Y_1 \\ Y_2 \\ Y_3 \end{bmatrix} = \begin{bmatrix} 1 & 0 & 0 & 0 \\ 0.82 & 1 & 0 & 0 \\ 0.67 & 0.82 & 1 & 0 \\ 0.55 & 0.67 & 0.82 & 1 \end{bmatrix} \cdot \begin{bmatrix} 0 \\ 10 \\ 1 \\ 0 \end{bmatrix}$$

and the result **Y** of the direct problem is

$$\begin{aligned} &Y_0 && = 0 \\ &Y_1 = 10 + 1 && = 11 \\ &Y_2 = 8.2 + 1 && = 9.2 \\ &Y_3 = 6.7 + 0.82 && = 7.52 \end{aligned}$$

or

$$\begin{bmatrix} Y_0 \\ Y_1 \\ Y_2 \\ Y_3 \end{bmatrix} = \begin{bmatrix} 0 \\ 11.0 \\ 9.20 \\ 7.52 \end{bmatrix}$$

As expected, the effect of $u_2 = 1$ is added to the result for the input $u_1 = 10$ from b.

Inverse problem solution of obtaining the estimated input $\mathbf{U}_{est}$ given **h** and the measured output **Y** for K=3, is obtained by forward substitution. Direct problem scalar equation permit the calculation by forward substitution from i=0 toward i=3:

$$Y_0 = h_0 \cdot u_0$$

gives

$$u_0 = Y_0 / h_0$$

$$Y_1 = h_1 \cdot u_0 + h_0 \cdot u_1 = Y_0 \cdot h_1 / h_0 + h_0 \cdot u_1$$

gives

$$u_1 = - Y_0 \cdot h_1 / h_0^2 + Y_1 / h_0$$

$$Y_2 = h_2 \cdot u_0 + h_1 \cdot u_1 + h_0 \cdot u_2 =$$

$$Y_0 \cdot h_2 / h_0 + Y_1 \cdot h_1 / h_0 - Y_0 \cdot h_1^2 / h_0^2 + h_0 \cdot u_2 =$$
$$Y_0 \cdot [h_2 / h_0 - h_1^2 / h_0^2] + Y_1 \cdot h_1 / h_0 + h_0 \cdot u_2$$

gives

$$u_2 = Y_0 \cdot [- h_2 / h_0^2 + h_1 / h_0^3] - Y_1 \cdot h_1 / h_0^2 + Y_2 / h_0$$
$$Y_3 = h_3 \cdot u_0 + h_2 \cdot u_1 + h_1 \cdot u_2 + h_0 \cdot u_3 =$$
$$Y_0 \cdot h_3/h_0 - Y_0 \cdot h_1 \cdot h_2 / h_0^2 + Y_1 \cdot h_2 / h_0 + Y_0 \cdot [-h_1 \cdot h_2 / h_0^2 + h_1^2 / h_0^3] -$$
$$Y_1 \cdot h_1^2 / h_0^2 + Y_2 \cdot h_1 / h_0 + h_0 \cdot u_3$$

or

$$Y_3 = Y_0 \cdot [h_3 / h_0 - h_1 \cdot h_2 / h_0^2 - h_1 \cdot h_2 / h_0^2 + h_1^2 / h_0^3] + Y_1[h_2 / h_0 -$$
$$h_1^2 / h_0^2] + Y_2 \cdot h_1 / h_0 + h_0 \cdot u_3$$

gives

$$u_3 = -Y_0 \cdot [h_3 / h_0^2 - 2h_1 \cdot h_2 / h_0^3 + h_1^2 / h_0^4] - Y_1 \cdot [h_2 / h_0^2 - h_1^2 / h_0^3] - Y_2 \cdot$$
$$h_1 / h_0^2 + Y_3 / h_0$$

In matrix form again results a lower triangular matrix [4 · 4]

$$\begin{bmatrix} u_0 \\ u_1 \\ u_2 \\ u_3 \end{bmatrix} =$$
$$\begin{bmatrix} 1/h_0 & 0 & 0 & 0 \\ -h_1/h_0^2 & 1/h_0 & 0 & 0 \\ -h_2/h_0^2 + h_1/h_0^3 & -h_1/h_0^2 & 1/h_0 & 0 \\ -h_3/h_0^2 + 2 \cdot h_1 h_2/h_0^3 - h_1^2/h_0^4 & -h_2/h_0^2 + h_1/h_0^3 & -h_1/h_0^2 & 1/h_0 \end{bmatrix} \cdot \begin{bmatrix} Y_0 \\ Y_1 \\ Y_2 \\ Y_3 \end{bmatrix}$$

This gives the inverse problem solution $\mathbf{U} = \mathbf{h}^{-1} \cdot \mathbf{Y}$, where

$$\mathbf{h}^1 =$$

$$\begin{bmatrix} 1/h_0 & 0 & 0 & 0 \\ -h_1/h_0^2 & 1/h_0 & 0 & 0 \\ -h_2/h_0^2 + h_1/h_0^3 & -h_1/h_0^2 & 1/h_0 & 0 \\ -h_3/h_0^2 + 2 \cdot h_1 h_2/h_0^3 - h_1^2/h_0^4 & -h_2/h_0^2 + h_1/h_0^3 & -h_1/h_0^2 & 1/h_0 \end{bmatrix}$$

Example 3.6 For a first order system $h(t) = e^{-at}\,1(t)$ with $a = 0.2$, and sampling time $T_s = 1$, obtain

a) h_i
b) scalar equations inverse problem solutions for u_i for i=1, 2, 3 and 4
c) estimate of input U given the output values $Y = [0\ 10\ \ 8.2\ \ 6.7]^T$.

a)

$$h_i = e^{-at_i} \text{ for } i = 0,1,2,3$$

gives

$$\begin{aligned} h_0 &= \exp\{-0.2 \cdot 0\} = 1 \\ h_1 &= \exp\{-0.2\} = 0.82 \\ h_2 &= \exp\{-0.4\} = 0.67 \\ h_3 &= \exp\{-0.6\} = 0.55 \end{aligned}$$

b)

$u_0 = Y_0 / h_0 = Y_0$

$u_1 = -\,Y_0 \cdot h_1 / h_0^2 + Y_1 / h_0 = -Y_0 \cdot 0.82 + Y_1$

$u_2 = Y_0 \cdot [-\,h_2 / h_0^2 + h_1 / h_0^3\,] - Y_1 \cdot h_1 / h_0^2 + Y_2 / h_0$

$\quad = Y_0 \cdot (-0.67 \ + 0.82\,) - Y_1 \cdot 0.82 + Y_2$

$u_3 = Y_0 \cdot [-\,h_3 / h_0^2 + 2 \cdot h_1 \cdot h_2 / h_0^3 - h_1^2 / h_0^4] + Y_1 \cdot [-\,h_2 / h_0^2 + h_1 / h_0^3] -$

$\quad\quad Y_2 \cdot h_1 / h_0^2 + Y_3 / h_0$

$\quad = Y_0 \cdot [-0.55 + 2 \cdot 0.82 \cdot 0.67 - 0.82^2] + Y_1 \cdot [-0.67 + 0.82] - Y_2 \cdot 0.82 + Y_3$

c) For $Y_0 = 0$, $Y_1 = 10$, $Y_2 = 8.2$ and $Y_3 = 6.7$

d)

$u_0 = Y_0 = 0$

$u_1 = -Y_0 \cdot 0.82 + Y_1 = 10$

$u_2 = Y_0 \cdot (-0.67 + 0.82) - Y_1 \cdot 0.82 + Y_2 = -8.2 + 8.2 = 0$

$u_3 = Y_0 \cdot [-0.55 + 2 \cdot 0.82 \cdot 0.67 - +0.82^2] + Y_1 \cdot [-0.67 + 0.82] - Y_2 \cdot 0.82 + Y_3$
$= 10 \cdot [-0.67 + 0.82] - 8.2 \cdot 0.82 + 6.7 = 0$

or in matrix form

$$\begin{bmatrix} 0 \\ 10 \\ 0 \\ 0 \end{bmatrix} = \begin{bmatrix} 1 & 0 & 0 & 0 \\ -0.82 & 1 & 0 & 0 \\ 0.15 & -0.82 & 1 & 0 \\ 1.22 & 0.15 & -0.82 & 1 \end{bmatrix} \cdot \begin{bmatrix} 0 \\ 10 \\ 8.2 \\ 6.7 \end{bmatrix}$$

The inverse problem solution **U**, given **Y** is, in this case of a non-singular square matrix $\mathbf{h}^{-1}$, the exact recovery of forward problem from Example 3.5.b. In the case of noisy signals **Y**, this exact recovery of **U** is not possible. In case of higher order sensors, with longer duration and non-square matrix $\mathbf{h}^{-1}$, such exact recovery is also not possible. Moreover, matrices $\mathbf{h}^{-1}$ can result in an ill-posed inverse problem.

For the linear time variant case,

$$Y_i = \sum_{k=i-K}^{i} h_{i,k} \cdot u_k$$

the solution for the inverse problem can also be calculated by forward substitution, but the K substitutions involve more computations because $h_{i,k} \neq h_{i-k}$ is dependent on actual tome values i and k, not only on the difference i - k.

B) $N_y < N_u$

In case that the length of the output vector N_y is shorter than the length of the input vector N_u, the inverse matrix cannot be computed and has to be replaced by the generalized inverse $\mathbf{h}^{-g}$. The inverse problem is in this case under-determined [7]

C) $N_y > N_u$
In case that the length of the output vector N_y is longer than the length of the input vector N_u, inverse matrix cannot be computed and a compromise solution can be computed using least squares method. Inverse problem is in this case over-determined [7].

These two last cases of non-square matrices **h** will be investigated later in this chapter. The analysis will be based on matrix form representation $\mathbf{Y} = \mathbf{h} \cdot \mathbf{U}$ which is not limited to lumped parameters system or to discrete Volterra equation case, when **h** is lower triangular. In what follows, **h** is not in general lower triangular, *i.e.* it corresponds to Fredholm equation and also to numerous other direct and inverse problems presented in this book.

3.3 Discrete Inverse Problems Solved by Matrix Inversion

3.3.1 *Types of Methods for Solving Inverse Problems*

Two types of methods for solving inverse problems will be presented:
-matrix inversion
-iterative methods using forward model.

The basic scheme of solving the discrete inverse problem using matrix inversion is shown in Fig. 3.1.

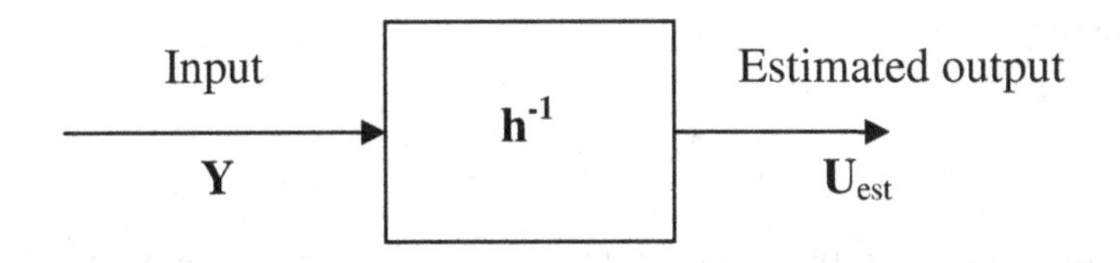

Fig. 3.1 The diagram of inverse problem solver using matrix inverse

Y, the output of the direct problem, becomes the input to the inverse problem and consists in the data from a system for which either the input **U** or the parameters **P** are unknown, while the model **f** (**P**, **U**) is assumed as a known forward problem

$$\mathbf{Y} = \mathbf{f}\,(\mathbf{P}, \mathbf{U})$$

The corresponding discrete matrix form of the forward problem is

$$\mathbf{Y} = \mathbf{h} \cdot \mathbf{U}$$

Inverse problems result from unknown input estimation [67]

$$\mathbf{U}_{est} = \mathbf{h}^{-g} \cdot \mathbf{Y}_m$$

or from parameters identification

$$\mathbf{h}_{est} = \mathbf{Y}_m \cdot \mathbf{U}^{-g}$$

Matrix inversion based solvers require generally the computation of a pseudo-inverse and, has to take into account that the data input to the inverse problem, $\mathbf{Y}_{m}$, are normally the result of noisy measurements, i.e. in fact that

$$\mathbf{Y}_m = \mathbf{Y} + \mathbf{w}$$

where **w** represents measurement noise and **Y** is the noiseless sensor output. In this case of noisy measurement data $\mathbf{Y}_{m}$, inverse problem becomes

$$\mathbf{U}_{est} = \mathbf{h}^{-g} \cdot (\mathbf{Y} + \mathbf{w})$$

Ill-posed inverse problem results when there is a very high amplification by some rows of $\mathbf{h}^{-g}$ of the significant variation of the adjacent elements of $\mathbf{Y}_{m}$, in particular due random noise **w**.

The basic scheme of solving the discrete inverse problem using iterative methods using forward model is shown in Fig. 3.2. It can be observed that solving the discrete inverse problem using matrix inversion is basically an open loop scheme, while iterative method using forward model is a closed loop error (negative feedback) based computation scheme. Iterative method using forward model is not affected by the difficulties resulting from the inversion of a matrix, particularly in the case of rank deficiency. Moreover, the aforementioned amplification of the noise effect in matrix inverse method is avoided in the iterative method. Being an iterative scheme, real-time applications might be however limited due to the computation duration cannot be predetermined for given ε.

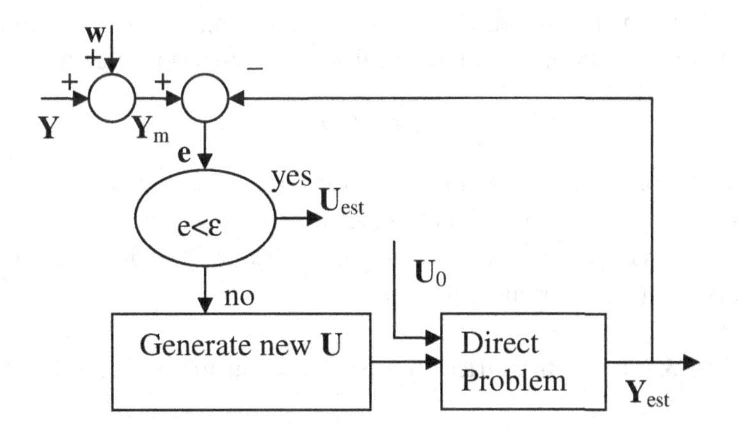

Fig. 3.2 Basic scheme of solving the discrete inverse problem using iterative methods

The concept behind the method for solving inverse problems using matrix inversion is the same as for open loop control, while for iterative methods using forward model is the same as for closed loop negative feedback control.

3.3.2 *Inverse and Pseudo-Inverse. MATLAB Solutions*

Discrete inverse problems result from approximating continuous system models but also directly from models of system monitoring.

In Ch. 3.1, both Fredholm and Volterra equations were approximated by systems of linear equations in discrete form leading to the matrix equation $\mathbf{Y} = \mathbf{h} \cdot \mathbf{U}$.

The solution of the above matrix equation is obtained using the inverse matrix $\mathbf{h}^{-1}$

$$\mathbf{U} = \mathbf{h}^{-1} \cdot \mathbf{Y}$$

if the inverse $\mathbf{h}^{-1}$ exists. Example 3.6 illustrates this type of problem.

Given that in monitoring, estimation, identification and control applications, matrix **h** is rarely square and non-singular, and the solution of the matrix equation uses the generalized or pseudo-inverse $\mathbf{h}^{-g}$

$$\mathbf{U}_{est} = \mathbf{h}^{-g} \cdot Y_m$$

MATLAB® contains numerous functions that solve inverse, pseudo-inverse and various over- and under-determined problems. In what follows, MATLAB examples illustrate solutions to discrete inverse problems in matrix formulation.

Example 3.7 Three **h** matrices are used for illustrating these functions [81]

$$1)\ \mathbf{A} = \begin{bmatrix} 1 & 3 \\ 5 & 7 \end{bmatrix}$$

$$2)\ \mathbf{B} = \begin{bmatrix} 1 & 3 \\ 2 & 6 \end{bmatrix}$$

$$3)\ \mathbf{C} = \begin{bmatrix} 1 & 3 & 4 \\ 5 & 7 & 2 \end{bmatrix}$$

Matrix **A** is square and non-singular, |**A**| = - 1, matrix **B** is square and singular, |**B**| = 0 and matrix **C** is non-square.

MATLAB function inv() gives

1) $\mathbf{A} = \begin{bmatrix} 1 & 3 \\ 5 & 7 \end{bmatrix}$

```
      inv(A)
-0.8750   0.3750
 0.6250  -0.1250
      pinv(A)
-0.8750   0.3750
 0.6250  -0.1250
```

For the square, non-singular matrix **A**, the inverse $\mathbf{A}^{-1}$ exists and is the same as the pseudo-inverse $\mathbf{A}^{-g}$.

2) $\mathbf{B} = \begin{bmatrix} 1 & 3 \\ 2 & 6 \end{bmatrix}$

```
>> inv(B)
Warning: Matrix is singular to working precision.
ans =
  Inf  Inf
  Inf  Inf
```

For this singular matrix (second row is the double of the first row) there is no inverse $\mathbf{B}^{-1}$.

The pseudo-inverse function gives however

```
pinv(B)
  0.0200  0.0400
  0.0600  0.1200
```

3) $\mathbf{C} = \begin{bmatrix} 1 & 3 & 4 \\ 5 & 7 & 2 \end{bmatrix}$

```
>> inv(C)
??? Error using ==> inv
Matrix must be square.
```

The non-square matrix **C** has no inverse $\mathbf{C}^{-1}$.

```
>> pinv(C)
  -0.1055    0.1101
  -0.0046    0.0917
   0.2798   -0.0963
```

The non-square matrix **C** has a pseudo-inverse $\mathbf{C}^{-g}$.
Further inside is given by the rank of these matrices

```
>> rank(A)
=2

>> rank (B)
 =1

>> rank(C)
= 2
```

It can be observed that matrices A $[2 \cdot 2]$ and **C** $[2 \cdot 3]$ have rank 2, while the singular matrix **B** $[2 \cdot 2]$ has the rank 1.

Also, the norm and condition number are needed for further developments in solving discrete inverse problems.

Euclidean norms $\|\mathbf{A}\|_2$ are

```
>> norm(A)
= 9.1231

>> norm(B)
=7.0711

>> norm(C)
=9.7366
```

Maximum norms $||A||_\infty$ are

```
>> norm(A, inf)
=12
```

```
>> norm(B, inf)
=8
```

```
>> norm(C, inf)
=14
```

Norms of absolute values $||\mathbf{A}||_1$ are

```
>> norm(A,1)
=10
```

```
>> norm(B,1)
=9
```

```
>> norm(C,1)
=10
```

More important in characterizing ill-posed inverse problems will be the condition number.

```
>> cond(A)
=10.4039
```

```
>> cond(B)
=1.2588e+016
```

```
>> cond(C)
=3.2104
```

It can be observed that for the singular matrix **B** $[2 \cdot 2]$ the rank(**B**) = 1 and the condition number is cond(**B**) = 1.2588e+016. The condition number is a better indicator of how close a square matrix is to being non-singular.

For example for the square matrix **B1** that resembles to **B** but the second row is not exactly the double of the first row

$$B1 = \begin{bmatrix} 1 & 3 \\ 2 & 6.0000000001 \end{bmatrix}$$

pinv(B1)
>> inv(B1)
= 1.0e+010 *
 5.999999503657816 -2.999999751778908
 -1.999999834519272 0.99999991725963
=
 5.9999e+010 -2.9999e+010
 -2.0000e+010 9.9998e+009

>> pinv(B1)
= 1.0e+010 *
 5.999853977167788 -2.999926988533896
 -1.999951325689263 0.999975662844632

>> rank(B1)
=2

>> cond(B1)
= 4.9999e+011

It can be observed that in this case the inverse $\mathbf{B1}^{-1}$ exists and is close to the values of $\mathbf{B1}^{-g}$, the rank of is 2, but the condition number cond(**B1**)= 4.9999e+011 is very large and closer to the large number of cond(**B**)=1.2588e+016, indicating that **B1** is closer to the singular matrix **B** than to the non-singular matrix **A**.

Further inside can be obtained from changing **B1** into **B2**

$$\mathbf{B2} = \begin{bmatrix} 1 & 3 \\ 2 & 6.0000000000000001 \end{bmatrix}$$

```
>> inv(b2)
Warning: Matrix is B2 is singular to working precision.
ans =
 Inf  Inf
 Inf  Inf

>> pinv(b2)
=
   0.0200   0.0400
   0.0600   0.1200
```

In this case, making **B2** has element (2, 2) very close to 6.0 from **B** gave a MATLAB result for pinv(b2) identical to the result from pinv(b1) while pinv(b1) gave a very different result. It can be observed, however, that within MATLAB working precision,

```
B*pinv(B)
=
   1.0000   0.0000
   0.0000   1.0000
```

results in an identity matrix [2 · 2], while for B1*pinv(B1)

```
B1*pinv(B1)
=
0.999984741210938   0.000015258789063
  0.000015258789063   1.000007629394531
```

results in a [2 · 2] matrix, that is approximately identity matrix in four digits approximation

```
>> B1*pinv(B1)
=
   1.0000   0.0000
   0.0000   1.0000
```

For B2*pinv(B2), the result is

```
B2*pinv(B2)
=
   0.2000   0.4000
   0.4000   0.8000
```

That is not an identity matrix. This shows that square matrices which are very close to singularity, to like **B1** and **B2**, or singular, like **B**, i.e. matrices with very high condition number, can result in significantly different pseudo-inverses.

Discrete inverse problems require more elaborate analysis using specific methods for over- and under-determined problems, condition number, and Singular Value Decomposition (SVD).

Example 3.8 The purpose is to obtain the discrete direct problem formulation for a DC motor.

Consider the DC motor from Example 1.5 for $\omega = d\theta / dt$

$$k_m \cdot i = J\frac{d\omega}{dt} - b \cdot \omega - T$$

$$u = L\frac{di}{dt} + R \cdot i + k_e \cdot \omega$$

For negligible L and b, (L = 0 and b = 0) and no load (T = 0)

$$J\frac{d\omega}{dt} = k_m \cdot i$$

$$u = R \cdot i - k_e \cdot \omega$$

For i obtained from the last equation,

$$i = \frac{u - k_e \cdot \omega}{R}$$

the torque equation becomes

$$J\frac{d\omega}{dt} = k_m \cdot \frac{u - k_e \cdot \omega}{R}$$

or

$$\dot{\omega} = -\frac{k_m \cdot k_e}{J \cdot R}\omega + \frac{k_m}{J \cdot R}u$$

or

$$\dot{\omega} = -\alpha \cdot \omega + \beta \cdot u$$

where

$$\alpha = -\frac{k_m \cdot k_e}{J \cdot R}$$

$$\beta = \frac{k_m}{J \cdot R}$$

The transfer function is

$$\frac{\omega(s)}{u(s)} = \frac{\beta}{1 + \alpha \cdot s}$$

Direct problem is given by the state space model, for

$$x_1 = \theta$$
$$x_2 = dx_1 / dt = d\theta/dt = \omega$$

such that

$$\begin{bmatrix} \dot{x}_1 \\ \dot{x}_2 \end{bmatrix} = \begin{bmatrix} 0 & 1 \\ 0 & -\alpha \end{bmatrix} \cdot \begin{bmatrix} x_1 \\ x_2 \end{bmatrix} + \begin{bmatrix} 0 \\ \beta \end{bmatrix} \cdot u$$

Discrete direct problem will be derived for a sampling period T and zero-order hold (sample and hold approach) [127].

Sampling frequency 1/T is chosen to satisfy sampling theorem requirement for the maximum frequency f_{max} of interest of the sampled signal

$$1 / T > 2 \cdot f_{max}$$

or, at limit

$$T = 1 / (2 \cdot f_{max})$$

Sampling interval of $\nu \cdot T$ duration is chosen larger than 2% settling time T_s. In this case, for the time constant α, $T_s = 4 \cdot \alpha$, such that

$$\nu \cdot T > 4 \cdot \alpha$$

or

$$\nu / (2 \cdot f_{max}) > 4 \cdot \alpha$$

or

$$\nu > 8 \cdot \frac{k_m \cdot k_e \cdot f_{max}}{J \cdot R}$$

Discrete time state representation is given by [70, 127]

$$x_{n+1} = \Phi \cdot x_n + \Gamma \cdot u_n \quad \text{for } n = 1,2,\ldots, \nu$$

where [70]

$$\Phi = e^{A \cdot T} \approx I + A \cdot T = \begin{bmatrix} 1 & T \\ 0 & 1 - \alpha \cdot T \end{bmatrix}$$

$$\Gamma = \int_0^{\lambda} e^{A \cdot \lambda} \cdot B \cdot d\lambda \approx (I + A \cdot T/2) \cdot T \cdot B = \begin{bmatrix} \beta \cdot T^2/2 \\ \beta(1 - \alpha \cdot T^2/2) \end{bmatrix}$$

By recursion

$$x_1 = \Phi \cdot x_0 + \Gamma \cdot u_0$$

$$x_2 = \Phi \cdot x_1 + \Gamma \cdot u_1 = \Phi^2 \cdot x_0 + \Phi \cdot \Gamma \cdot u_0 + \Gamma \cdot u_1$$

$$\cdots\cdots\cdots$$

$$x_\nu = \Phi^\nu \cdot x_0 + \Phi^{\nu-1} \cdot \Gamma \cdot u_0 + \ldots \Gamma \cdot u_{\nu-1}$$

or

$$x_\nu = \Phi^\nu x_0 + \sum_{k=0}^{\nu-1} \Phi^{\nu-1-k} \cdot \Gamma \cdot u_k$$

The last equation is the discrete time convolution integral of a LTI Volterra equation.

In matrix form, for zero initial state $\mathbf{x}_0 = \mathbf{0}$, the direct problem $\mathbf{x} = \mathbf{h} \cdot \mathbf{U}$ in discrete form is

$$\begin{bmatrix} x_1 \\ x_2 \\ \cdot \\ \cdot \\ x_\nu \end{bmatrix} = \begin{bmatrix} \Gamma & 0 & . & . & 0 \\ \Phi \cdot \Gamma & \Gamma & . & . & 0 \\ \cdot & \cdot & . & . & .. \\ \cdot & \cdot & . & . & . \\ \Phi^{\nu-1} \cdot \Gamma & \Phi^{\nu-2} \cdot \Gamma & . & . & \Gamma \end{bmatrix} \cdot \begin{bmatrix} u_0 \\ u_1 \\ \cdot \\ \cdot \\ u_{\nu-1} \end{bmatrix}$$

This lower triangular h square matrix permits to formulate the inverse problem as

$$\mathbf{U} = \mathbf{h}^{-1} \cdot \mathbf{x}$$

This inverse problem allows to recover the sequence of commands u_0, u_1,..., u_{v-1} that resulted in state values x_1, x_2,..., x_v. MATLAB function c2d converts the continuous time equation

$$d\mathbf{X}(t) / dt = \mathbf{A} \cdot \mathbf{X}(t) + \mathbf{B} \cdot \mathbf{u}(t)$$

into discrete time equation

$$\mathbf{x} = \mathbf{h} \cdot \mathbf{U}$$

for zero initial conditions.

3.3.3 *Over-Determined and Under-Determined Problems*

Assume the matrix equation **Y= h · U** where **Y** $[N_y \cdot 1]$, **h** $[N_y \cdot N_u]$ and **U** $[N_u \cdot 1]$.

This equation is

a) over-determined for $N_y > N_u$, i.e. more data values than unknowns
b) even-determined for $N_y = N_u$ and non-singular matrix h
c) under determined for $N_y < N_u$ i.e. fewer data values than unknowns

Over-determined matrix equations for $N_y > N_u$ are solved using Least-Squares Solution (LSS). This solution results from minimizing the Euclidean norm of the errors, a least squares error solution [28, 65]

$$\min S(\mathbf{U}) = \min\{(\mathbf{Y}_m - \mathbf{h} \cdot \mathbf{U})^T \cdot (\mathbf{Y}_m - \mathbf{h} \cdot \mathbf{U})\}$$

The solution is obtained from vanishing partial derivatives

$$\delta S / \delta \mathbf{U} = -2 \cdot \mathbf{h}^T \cdot \mathbf{Y}_m + 2 \cdot \mathbf{h}^T \cdot \mathbf{h} \cdot \mathbf{U} = 0$$

If the rank $(\mathbf{h}) = m$, $\mathbf{h} \cdot \mathbf{h}^T$ is invertible and the solution for $\mathbf{U}$ is

$$\mathbf{U}_{est} = (\mathbf{h}^T \cdot \mathbf{h})^{-1} \cdot \mathbf{h}^T \cdot \mathbf{Y}_m$$

Even-determined solution for an invertible matrix $\mathbf{h}$, $N_y = N_u$, is given by

$$\mathbf{U}_{est} = \mathbf{h}^{-1} \cdot \mathbf{Y}_m$$

Under-determined for $N_y < N_u$ is solved using a minimum Euclidean length $\mathbf{U}^T \cdot \mathbf{U}$ solution subject to the constraint $\mathbf{Y}_m - \mathbf{h} \cdot \mathbf{U} = 0$, i.e. a Minimum Length Solution (MLS)

$$\min S(\mathbf{U}) = \min\{\mathbf{U}_m^T \cdot \mathbf{U} + \boldsymbol{\lambda}^T \cdot (\mathbf{Y}_m - \mathbf{h} \cdot \mathbf{U})\}$$

where $\boldsymbol{\lambda}$ is a Lagrange multiplier.

The solution is obtained from vanishing partial derivatives

$$\delta S / \delta \mathbf{U} = 2 \cdot \mathbf{U} - \mathbf{h}^T \cdot \boldsymbol{\lambda} = 0$$

$$\delta S / \delta \boldsymbol{\lambda} = \mathbf{Y}_m - \mathbf{h} \cdot \mathbf{U} = 0$$

From first equation

$$2 \cdot \mathbf{h} \cdot \mathbf{U} = \mathbf{h} \cdot \mathbf{h}^T \cdot \boldsymbol{\lambda}$$

Using second equation

$$2 \cdot \mathbf{Y}_m = 2 \cdot \mathbf{h} \cdot \mathbf{U} = \mathbf{h} \cdot \mathbf{h}^T \cdot \boldsymbol{\lambda}$$

If the rank ($\mathbf{h}$) = N_y, $\mathbf{h} \cdot \mathbf{h}^T$ is invertible and the solution for $\boldsymbol{\lambda}$ is

$$\boldsymbol{\lambda} = 2 \cdot (\mathbf{h} \cdot \mathbf{h}^T)^{-1} \cdot \mathbf{Y}_m$$

and the MLS solution for x is obtained replacing $\boldsymbol{\lambda}$ in the first vanishing partial derivative

$$2 \cdot \mathbf{U} = \mathbf{h}^T \cdot \boldsymbol{\lambda} = \mathbf{h}^T \cdot 2 \cdot (\mathbf{h} \cdot \mathbf{h}^T)^{-1} \cdot \mathbf{Y}_m$$

or

$$\mathbf{U}_{est} = \mathbf{h}^T \cdot (\mathbf{h} \cdot \mathbf{h}^T)^{-1} \cdot \mathbf{Y}_m$$

where the generalized inverse for the matrix $\mathbf{h}$ $[N_y \cdot N_u]$ is

$$\mathbf{h}^{-g} = \mathbf{h}^T \cdot (\mathbf{h} \cdot \mathbf{h}^T)^{-1}$$

MATLAB functions h\Y can solve over-determined, even-determined and under-determined matrix equations.

Example 3.9 For $\mathbf{Y} = \mathbf{h} \cdot \mathbf{U}$ in over-determined case $N_y > N_u$, for the matrix $\mathbf{h}$ [3 · 2] and $\mathbf{Y}$ [3 · 1], i.e. for $(N_y = 3) > (N_u = 2)$, rank(h) = 2, obtain the least squares solution $\mathbf{U}$ [2 · 1], $\mathbf{U} = \mathbf{h}^{-g} \cdot \mathbf{Y}$
where

```
>> h=[1 0;2 1;2 2]
h =
    1  0
    2  1
    2  2

>> rank(h)
=2
```

```
>> Y=[1;2;-4]
Y =
   1
   2
  -4
>> h\Y

ans =
    2.3333
   -4.0000
>>
```

This least square solution for **U** [2 · 1] minimizes total Euclidean error.

Example 3.10 For **Y** = **h** · **U** in over-determined case $N_y > N_u$, for the matrix **h** [3 · 2] and **Y** [3 · 1], i.e. for (N_y = 3) > (N_u = 2), rank(h) =2, obtain the least squares solution **U** [2 · 1], **U** = $\mathbf{h}^{-g}$ · **Y** where
h=[2 3;2 3;2 3]

```
h =
   2   3
   2   3
   2   3
>> rank(h)
ans =
   1
>> h\Y
Warning: Rank deficient, rank = 1,

ans =
       0
   -0.1111
```

Matrix h [3x2] has Rank deficient rank(h) = 1 and the solution for **U**[2 · 1] contains only one nonzero value -0.1111.

Example 3.11 For **Y** = **h** · **U** in underdetermined case $N_y < N_u$, for the matrix **h** [2 · 3] and **Y** [2 · 1], i.e. for $(N_y = 2) < (N_u = 3)$, rank(h) = 2, obtain the least squares solution **U** [3 · 1], $\mathbf{U} = \mathbf{h}^{-g} \cdot \mathbf{Y}$
where

```
>> h=[1 0 3;2 1 4]
h =
   1  0  3
   2  1  4
>> rank (h)
ans =2
>> Y=[4;5]
Y =
   4
   5
>> h\Y
ans =
   0
  -0.3333
   1.3333
>>
```

i.e.

$$U = \begin{bmatrix} 0 \\ -1/3 \\ 4/3 \end{bmatrix}$$

For a rank(h) = 2, only two nonzero results are obtained for **U** [3 · 1]. In fact, there are infinite solutions **U**(t) for arbitrary value of t but corresponding to the above solution for t = 0, [81]

$$\mathbf{h} \cdot \begin{bmatrix} t \\ -1/3 + a \cdot t \\ 4/3 + b \cdot t \end{bmatrix} = \mathbf{Y}$$

The parameters a and b can be calculated from the equation **h** · **U**(t) = **Y**

$$\begin{bmatrix} 1 & 0 & 3 \\ 2 & 1 & 4 \end{bmatrix} \begin{bmatrix} t \\ -1/3 + a \cdot t \\ 4/3 + b \cdot t \end{bmatrix} = \begin{bmatrix} 4 \\ 5 \end{bmatrix}$$

which gives two scalar equations with unknowns a and b

$$t + 4 + 3 \cdot b \cdot t = 4$$
$$2 \cdot t - 1/3 + a \cdot t + 16/3 + 4 \cdot b \cdot t = 5$$

or

$$(1 + 3 \cdot b) \cdot t = 0$$
$$(2 + a + 4 \cdot b) \cdot t = 0$$

For the general case $t \neq 0$, these equations are verified for the solutions

$$a = -2/3$$
$$b = -1/3$$

The infinite number of solutions U(t) for this under-determined problem is given for t taking any real number value in

$$\mathbf{U}(t) = \begin{bmatrix} t \\ -1/3 - (2/3) \cdot t \\ 4/3 - (1/3) \cdot t \end{bmatrix}$$

Example 3.12 For $\mathbf{Y} = \mathbf{h} \cdot \mathbf{U}$ in over-determined case $N_y > N_u$, for the matrix **h** $[6 \cdot 2]$ and **Y** $[6 \cdot 1]$, i.e. for $(N_y = 6) > (N_u = 2)$, rank(h) = 2, obtain the least squares solution **U** $[2 \cdot 1]$, $\mathbf{U} = \mathbf{h}^{-g} \cdot \mathbf{Y}$, for the example of experimental data for the measured resistance versus temperature Ch. 3.1, that consists of $N_y = 6$ pairs of values

i	1	2	3	4	5	6
R_i [Ω]	600	627	651	682	701	732
T_i [^{0}C]	24	25.2	27	28.7	29	29.8

There are $N_u = 2$ unknown parameters, a_0 and a_1 of the assumed linear direct problem model

$$R = a_0 + a_1 \cdot T$$

For the over-determined matrix equation

$$\mathbf{Y} = \mathbf{h} \cdot \mathbf{U}$$

the vector of unknowns is

$$\mathbf{U} = \begin{bmatrix} a_0 \\ a_1 \end{bmatrix}$$

For **Y** [6 · 1], **h** [6 · 2] and **U** [2 · 1], the matrix equation becomes

$$\begin{bmatrix} 600 \\ 627 \\ 651 \\ 682 \\ 701 \\ 732 \end{bmatrix} = \begin{bmatrix} 1 & 24.0 \\ 1 & 25.2 \\ 1 & 27.0 \\ 1 & 28.7 \\ 1 & 29.0 \\ 1 & 29.8 \end{bmatrix} \begin{bmatrix} a_0 \\ a_1 \end{bmatrix}$$

MATLAB program gives

```
>> h=[1 24.0;1 25.2;1 27.0;1 28.7;1 29.0;1 29.8]
h =
   1.0000  24.0000
   1.0000  25.2000
```

```
   1.0000  27.0000
   1.0000  28.7000
   1.0000  29.0000
   1.0000  29.8000
>> Y=[600;627;651;682;701;732]
Y =
  600
  627
  651
  682
  701
  732
>> h\Y
ans =
  97.9112
  20.8035
>>
```

The over-determined inverse problem solution using least squares errors method resulted in

$$\mathbf{U} = \mathbf{h}^{-g} \cdot \mathbf{Y} = \begin{bmatrix} a_0 \\ a_1 \end{bmatrix} = \begin{bmatrix} 97.9112 \\ 20.8035 \end{bmatrix}$$

Example 3.13 For $\mathbf{Y} = \mathbf{h} \cdot \mathbf{U}$ for the matrix $\mathbf{h}$ [4 · 2] and $\mathbf{Y}$ [4 · 1], *i.e.* for (N_y = 4) > (N_u = 2), rank(h) =2, obtain the least squares solution $\mathbf{U}$ [2 · 1], $\mathbf{U} = \mathbf{h}^{-g} \cdot \mathbf{Y}$, in over-determined case $N_y > N_u$, but of deficient rank of h. Assume N_y = 4 pairs of values, which correspond to the case of an insignificant variation of the resistance value versus temperature:

i	1	2	3	4
R_i [Ω]	600	600.0000000001	600.0000000002	599.99999999999
T_i [^{0}C]	24	25.2	27	28.7

There are $N_u = 2$ unknown parameters, a_0 and a_1 of the assumed linear direct problem model

$$R = a_0 + a_1 \cdot T$$

The over-determined matrix equation is $\mathbf{Y} = \mathbf{h} \cdot \mathbf{U}$, where the vector of unknowns is

$$\mathbf{U} = \begin{bmatrix} a_0 \\ a_1 \end{bmatrix}$$

For $\mathbf{Y}$ [4 · 1], $\mathbf{h}$ [4 · 2] and $\mathbf{U}$ [2 · 1], the matrix equation becomes

$$\begin{bmatrix} 600 \\ 600.0000000001 \\ 600.0000000001 \\ 599.99999999999 \end{bmatrix} = \begin{bmatrix} 1 & 24 \\ 1 & 25.2 \\ 1 & 27 \\ 1 & 28.7 \end{bmatrix} \begin{bmatrix} a_0 \\ a_1 \end{bmatrix}$$

MATLAB program gives

```
>> Y=[600;600.0000000001;600.0000000001;599.99999999999]
Y =
 600.0000
 600.0000
 600.0000
 600.0000
>> h=[1 24;1 25.2;1 27;1 28.7]
h =
   1.0000  24.0000
   1.0000  25.2000
   1.0000  27.0000
   1.0000  28.7000
```

```
>> h\Y
ans =
  600.0000
   -0.0000
```

In fact, the result is $a_0 \approx 600$ and $a_1 \approx 0$.

The answer shows that in fact the resistance is practically constant for various temperature values in this case and the linear approximation is a constant

$$R = a_0 + a_1 \cdot T = 600.00$$

MATLAB gives

```
>> rank(h)
ans = 2
>> cond(h)
ans =387.8996
```

The rank equal to 2 would indicate that the matrix h would give a nonzero answer to both unknowns a_0 and a_1, but the condition number is high and indicates that the rank might be closer to 1.

In practice inverse problems can become mixed determined (partly over-determined, partly underdetermined) and generic solutions in this case are:

-Singular Value Decomposition (SVD) method
-Damped LS method
-Regularization method

These methods also address specific problems of ill-conditioning in inverse problems.

3.3.4 *SVD Method*

SVD method permits to determine the null-space of the forward problem $\mathbf{Y} = \mathbf{h} \cdot \mathbf{U}$ ($\mathbf{Y}$ $[N_y \cdot 1]$, $\mathbf{h}$ $[N_y \cdot N_u]$ and $\mathbf{U}$ $[N_u \cdot 1]$), when the matrix $\mathbf{h}$ is mapping the unknown vector **U** into the output vector **Y** with a zero subset. Such a zero subset of **Y**, the null space, cannot contribute to recover the unknown vector **U**.

SVD permits the calculation of a generalized inverse $\mathbf{h}^{-g}$ by retaining only singular values of **h** that do not result in ill-conditioning when calculating the estimation $\mathbf{U}_{ext}$ given $\mathbf{Y}_m$

$$\mathbf{U}_{ext} = \mathbf{h}^{-g} \cdot \mathbf{Y}_m$$

A non-square matrix **h** for $N_y > N_u$, (over-determined) and rank(h) = r $\leq N_u$, can be factored into

$$\mathbf{h} = \mathbf{u} \cdot \mathbf{\Lambda} \cdot \mathbf{v}^T$$

where:

$\mathbf{\Lambda}$ $[N_y \cdot N_u]$ is a matrix with all elements equal to zero except for diagonal values of the top left r · r part containing in decreasing order the r non-zero eigenvalues of $\mathbf{h} \cdot \mathbf{h}^T$ $[N_y \cdot N_y]$ or $\mathbf{h}^T \cdot \mathbf{h}$ $[N_u \cdot N_u]$

u $[N_y \cdot N_y]$ is an orthogonal matrix with the columns the eigenvectors of $\mathbf{h} \cdot \mathbf{h}^T$ calculating in the decreasing order of eigenvalues

v $[N_u \cdot N_u]$ is an orthogonal matrix with the columns the eigenvectors of $\mathbf{h}^T \cdot \mathbf{h}$ calculated in the decreasing order of eigenvalues.

The calculation by SVD method of $\mathbf{h}^{-g}$ $[N_u \cdot N_y]$ retains from $\mathbf{\Lambda}$ $[N_y \cdot N_u]$ in only $p \leq \min [N_y, N_u]$ eigenvalues that permit the avoidance of ill-conditioning by removing zero and very small valued eigenvalues and retaining only decreasing eigenvalues with no very low values Λ_1, Λ_2, ..., Λ_p, i.e. $\mathbf{\Lambda}_p$ $[p \cdot p]$, and calculates the corresponding $\mathbf{v}_p$ $[N_u \cdot p]$ and $\mathbf{u}_p$ $[N_y \cdot p]$ and $\mathbf{u}_p^T$ $[p \cdot N_y]$. The pseudo (or generalized) inverse is then calculated as follows

$$\mathbf{h}_p^{-g} = \mathbf{v}_p \cdot \mathbf{\Lambda}_p^{-1} \cdot \mathbf{u}_p^T$$

The elimination of zero and very low eigenvalues corresponds to a reduced order dynamic model that avoids ill-conditioning in inverse problem solving.

In this case

$$\mathbf{U}_{ext} = \mathbf{h}_p^{-g} \cdot \mathbf{Y}_m$$

gives

$$\mathbf{U}_{ext} = \mathbf{v}_p \cdot \mathbf{\Lambda}_p^{-1} \cdot \mathbf{u}_p^T \cdot \mathbf{Y}_m$$

where $\mathbf{\Lambda}_p$ is a diagonal matrix

$$\begin{bmatrix} \Lambda_1 & & & & \\ & \Lambda_2 & & & \\ & & \cdot & & \\ & & & \cdot & \\ & & & & \Lambda_p \end{bmatrix}$$

and $\mathbf{\Lambda}_p^{-1}$ is also a diagonal matrix and can be written as follows

$$\begin{bmatrix} 1/\Lambda_1 & & & & \\ & 1/\Lambda_2 & & & \\ & & \cdot & & \\ & & & \cdot & \\ & & & & 1/\Lambda_p \end{bmatrix}$$

such that $\mathbf{x}_{ext}$ can be expanded as follows [67]

$$\mathbf{U}_{est} = \sum_{i=1}^{p} \frac{1}{\Lambda_i} \mathbf{u}_i^T \cdot \mathbf{Y}_m \cdot \mathbf{v}_i$$

This shows that in order to obtain the inverse problem SVD solution the measurement values $\mathbf{Y}_m$ are divided by the decreasing eigenvalues $\Lambda_1 > \Lambda_2 >, \ldots, > \Lambda_p$. As a result, smaller eigenvalues in the denominator become high multiplication factors for $\mathbf{Y}_m = \mathbf{Y} + \mathbf{w}$ and their effect might become unacceptable when the errors w are more important then the contribution of **Y**, the usual case for high frequency components. As a result of removing smaller $\Lambda < \Lambda_p$ this unacceptable effect is avoided, making the inverse problem well-conditioned.

MATLAB examples will illustrate the use of SVD.

Example 3.14 SVD solution for over-determined case $N_y > N_u$
For the matrix **h** $[3 \cdot 2]$ and **Y** $[3 \cdot 1]$, *i.e.* for $(N_y = 3) > (N_u = 2)$, rank(h) = 2, obtain the SVD solution for **U** $[2 \cdot 1]$, $\mathbf{U} = \mathbf{h}^{-g} \cdot \mathbf{Y}$

```
>> h=[1 0;2 1;2 2]
h =
    1   0
    2   1
    2   2
>> [ua,Sa,va]=svd(h)
ua =
  -0.2222   0.7115   0.6667
  -0.6047   0.4358  -0.6667
  -0.7648  -0.5513   0.3333
Sa =
   3.6503        0
        0   0.8219
        0        0
va =
  -0.8112   0.5847
  -0.5847  -0.8112
```

Retaining decreasing eigenvalues $\Lambda_1 = 3.6503$ and $\Lambda_2 = 0.8219$ for $p = N_u = 2$, Λ_p, *i.e.* $\mathbf{\Lambda}_p (p \cdot p)$

becomes in the MATLAB notation

```
Sa=Sp
3.6503        0
     0   0.8219
```

or

$$\mathbf{S}_p = \begin{bmatrix} 3.6503 & 0 \\ 0 & 0.219 \end{bmatrix}$$

and

$\mathbf{Sp}^{-1}$

1/3.6503 0
0 1/0.8219

The calculation of the corresponding $\mathbf{v}_p$ $[N_u \cdot p]$ and $\mathbf{u}_p$ $[N_y \cdot p]$ and $\mathbf{u}_p^T$ $[p \cdot N_y]$ gives

Up =
-0.2222 0.7115
-0.6047 0.4358
-0.7648 -0.5513

and

u_p^T =
-0.2222 -0.6047 -0.7648
0.7115 0.4358 -0.5513

while $\mathbf{v}_p$ $(N_u \cdot p)$ remains the same for $p = N_u = 2$

vp =
-0.8112 0.5847
-0.5847 -0.8112

Finally

$$\mathbf{h}^{-g} = \mathbf{v}_p \cdot \mathbf{\Lambda}_p^{-1} \cdot \mathbf{u}_p^T$$

MATLAB gives

```
>> vp=[-0.8112 0.5847;-05847 -0.8112]
vp =
 1.0e+003 *
  -0.0008   0.0006
  -5.8470  -0.0008
>> upT=[-0.2222 -0.6047 -0.7648;0.7115 0.4358 -0.5513]
upT =
  -0.2222  -0.6047  -0.7648
   0.7115   0.4358  -0.5513
>> Spinv=[1/3.6503 0;0 1/0.8219]
Spinv =
   0.2740        0
        0   1.2167
>> hinv=vp*Spinv*upT
hinv =
 1.0e+003 *
  0.000555540524813  0.000444409759338  -0.000222234836570
  0.355214646288678  0.968169961176620  1.225590173846834
= 5.5554e-001  4.4441e-001 -2.2223e-001
   3.5521e+002  9.6817e+002  1.2256e+003
>> Ym=[1;2;-4]
Ym =
    1
    2
   -4
>> Uest=hinv*Ym
Uest =
 1.0e+003 *
   0.0023
  -2.6108 =
```

or

$$\mathbf{U}_{\text{est}} = \begin{bmatrix} 2.3333 \\ -2610.8 \end{bmatrix}$$

The SVD result has the same value for the first element, 0.2333 and different value for the second element, -2.6108, versus -4.000, for the h\Y earlier calculation. The difference is due to the use of different algorithms and approximations.

3.3.5 *Damped LS Solution*

Damped LS (DLS) method is a combination of the LS solution with minimum length solution resulting from

$$\min S(\mathbf{U}) = \min\{(\mathbf{Y}_m - \mathbf{h} \cdot \mathbf{U})^T \cdot (\mathbf{Y}_m - \mathbf{h} \cdot \mathbf{U}) + \boldsymbol{\lambda} \cdot \mathbf{U}^T \cdot \mathbf{U}\}$$

Vanishing partial derivative gives

$$\delta S/\delta \mathbf{U} = -2 \cdot \mathbf{h}^T \cdot \mathbf{Y}_m + 2 \cdot \mathbf{h}^T \cdot \mathbf{h} \cdot \mathbf{U} + 2\,\boldsymbol{\lambda} \cdot \mathbf{U} = 0$$

In case that $\mathbf{h} \cdot \mathbf{h}^T + \boldsymbol{\lambda} \cdot \mathbf{I}$ is non-singular,

$$\mathbf{U}_{est} = (\mathbf{h}^T \cdot \mathbf{h} + \boldsymbol{\lambda} \cdot \mathbf{I})^{-1} \cdot \mathbf{h}^T \cdot \mathbf{Y}_m$$

As expected, for $\boldsymbol{\lambda}=0$, damped LS solution becomes identical to LS solution

$$\mathbf{U}_{est} = (\mathbf{h}^T \cdot \mathbf{h})^{-1} \cdot \mathbf{h}^T \cdot \mathbf{Y}_m$$

3.3.6 *Regularization Method. Regularized LSS*

Regularization method, proposed initially by Tikhonov, is a combination of the LS solution with a priori information about how "smooth" the solution has to be and leads to a stable solution for the inverse problem. A priori information does not come from the data measurement information contained in $\mathbf{Y}_m$ vector [45, 47, 63, 67]. A priori information or belief is contained in the regularization matrix $\mathbf{R}$ $[k \cdot N_u]$ that multiplies $\mathbf{U}$ $[N_u \cdot 1]$ in

$$\min S(\mathbf{U}) = \min\{(\mathbf{Y}_m - \mathbf{h} \cdot \mathbf{U})^T \cdot (\mathbf{Y}_m - \mathbf{h} \cdot \mathbf{U}) + \boldsymbol{\lambda} \cdot (\mathbf{R} \cdot \mathbf{U})^T \cdot (\mathbf{R} \cdot \mathbf{U})\}$$

or

$$\min S(\mathbf{U}) = \min\{(\mathbf{Y}_m - \mathbf{h} \cdot \mathbf{U})^T \cdot (\mathbf{Y}_m - \mathbf{h} \cdot \mathbf{U}) + \boldsymbol{\lambda} \cdot (\mathbf{U}^T \cdot \mathbf{R}^T \cdot \mathbf{R} \cdot \mathbf{U})\}$$

Vanishing partial derivative gives

$$\delta S/\delta \mathbf{U} = -2 \cdot \mathbf{h}^T \cdot \mathbf{Y}_m + 2 \cdot \mathbf{h}^T \cdot \mathbf{h} \cdot \mathbf{U} + 2\boldsymbol{\lambda} \cdot \mathbf{R}^T \cdot \mathbf{R} \cdot \mathbf{U} = 0$$

Given any non zero k of $\mathbf{R}$ $[k \cdot N_u]$, the matrix $\mathbf{R}^T \cdot \mathbf{R}$ is $[N_u \cdot N_u]$ is square.

In case that $\mathbf{h}^T \cdot \mathbf{h} + \boldsymbol{\lambda} \cdot \mathbf{R}^T \cdot \mathbf{R}$ is non-singular, by inversion gives the regularized LS solution

$$\mathbf{U}_{est} = (\mathbf{h}^T \cdot \mathbf{h} + \boldsymbol{\lambda} \cdot \mathbf{R}^T \cdot \mathbf{R})^{-1} \cdot \mathbf{h}^T \cdot \mathbf{Y}_m$$

As expected, for $\mathbf{R} = \mathbf{I}$, regularized LS solution becomes identical to the damped LS solution

$$\mathbf{U}_{est} = (\mathbf{h}^T \cdot \mathbf{h} + \boldsymbol{\lambda} \cdot \mathbf{I})^{-1} \cdot \mathbf{h}^T \cdot Y_m$$

and for $\boldsymbol{\lambda}=0$, the regularized LS solution becomes identical to LS solution

$$\mathbf{U}_{est} = (\mathbf{h}^T \cdot \mathbf{h})^{-1} \cdot \mathbf{h}^T \cdot \mathbf{Y}_m$$

Regularization matrix $\mathbf{R}$ $[k \cdot N_u]$ reflects a priori information about the solution U, while a posteriori information is contained in $\mathbf{Y}_m$.

Examples of a priori information about the solution vector $\mathbf{U}$ or sequence of adjacent scalar values U_{k-1}, U_k, … for k = 1, 2,…, K are [49, 67]:

quasi constant, i.e. minimum change of $U_k - U_{k-1}$, for $k = 1, 2, \ldots, N_u - 1$
linear variation, i.e. minimum change of $(U_{k+1} - 2\,U_k + U_{k-1})$, for $k = 1, 2, \ldots, N_u - 2$
quadratic variation, i.e. minimum change of $(U_{k+2} - 3\,U_{k+1} + 3\,U_k - U_{k-1})$, for $k = 1, 2, \ldots, N_u - 3$ etc.

The effect of a priori information or belief about the sequence of adjacent scalar values U_{k-1}, U_k, … for k = 1, 2,…, K of the vector $\mathbf{U}$ $[K \cdot N_u]$ is to reduce the variability inherent to ill-posed inverse problems due to potential high amplification of errors. They work by replacing current value U_k by a weighted sum of adjacent values ..., U_{k-1}, U_k, U_{k+1},....

These three cases lead to the following non-square regularization matrices:

a) quasi constant, **R** $[(N_u - 1) \cdot N_u]$

$$\mathbf{R} = \begin{bmatrix} -1 & 1 & 0 & . & 0 \\ 0 & -1 & 1 & . & 0 \\ . & . & . & . & . \\ 0 & . & -1 & 1 & 0 \\ 0 & 0 & 0 & -1 & 1 \end{bmatrix}$$

b) linear variation, **R** $[(N_u - 2) \cdot N_u]$

$$\mathbf{R} = \begin{bmatrix} -1 & 2 & -1 & 0 & . & . & 0 \\ 0 & -1 & 2 & -1 & . & . & 0 \\ 0 & 0 & -1 & 2 & . & . & 0 \\ . & . & . & . & . & . & . \\ 0 & . & -1 & 2 & -1 & . & 0 \\ 0 & . & . & -1 & 2 & -1 & 0 \\ 0 & . & . & . & -1 & 2 & -1 \end{bmatrix}$$

c) quadratic variation, **R** $[(N_u - 3) \cdot N_u]$

$$\mathbf{R} = \begin{bmatrix} -1 & 3 & -3 & 1 & . & . & 0 \\ 0 & -1 & 3 & -3 & . & . & 0 \\ 0 & 0 & -1 & 3 & . & . & 0 \\ . & . & . & . & . & . & . \\ 0 & . & 3 & -3 & 1 & . & 0 \\ 0 & . & . & 3 & -3 & 1 & 0 \\ 0 & . & . & -1 & 3 & -3 & 1 \end{bmatrix}$$

The calculation of the solution requires not only the regularization matrix **R**, for example as defined above, but also a value for the Lagrange amplifier $\boldsymbol{\lambda}$ for

$$\mathbf{U}_{est} = (\mathbf{h}^T \cdot \mathbf{h} + \boldsymbol{\lambda} \cdot \mathbf{R}^T \cdot \mathbf{R})^{-1} \cdot \mathbf{h}^T \cdot \mathbf{Y}_m$$

A first value for a scalar λ is suggested to be [49]

$$\lambda = \mathrm{Tr}\,(\mathbf{h}^T \cdot \mathbf{h}) \,/\, \mathrm{Tr}\,(\mathbf{h})$$

using the trace of these matrices, i.e. the sum of diagonal elements. In obtaining a suitable value for λ, we have to take into account that low values in the minimization of S

$$\min S(\mathbf{U}) = \min\{(\mathbf{Y_m} - \mathbf{h} \cdot \mathbf{U})^T \cdot (\mathbf{Y}_m - \mathbf{h} \cdot \mathbf{U}) + \lambda \cdot (\mathbf{R} \cdot \mathbf{U})^T \cdot (\mathbf{R} \cdot \mathbf{U})\}$$

favour the part

$$(\mathbf{Y}_m - \mathbf{h} \cdot \mathbf{U})^T \cdot (\mathbf{Y}_m - \mathbf{h} \cdot \mathbf{U})$$

while high values favour the part

$$(\mathbf{R} \cdot \mathbf{U})^T \cdot (\mathbf{R} \cdot \mathbf{U})$$

Example 3.15 The solution for over-determined case $N_y > N_u$ for $\mathbf{Y} = \mathbf{h} \cdot \mathbf{U}$. For the matrix **h** [3 · 2] and **Y** [3 · 1], *i.e.* for $(N_y = 3) > (N_u = 2)$, rank(h) = 2, MATLAB gave the following least squares solution for **U** [2 · 1], $\mathbf{U} = \mathbf{h}^{-g} \cdot \mathbf{Y}$

```
>> h=[1 0;2 1;2 2]
h =
    1   0
    2   1
    2   2
>> rank(h)
ans =2
>> Y=[1;2;-4]
Y =
    1
    2
   -4
>> h\Y
ans =
   2.3333
  -4.0000
>>
```

Regularization solution

$$\mathbf{U}_{est} = (\mathbf{h}^T \cdot \mathbf{h} + \lambda \cdot \mathbf{R}^T \cdot \mathbf{R})^{-1} \cdot \mathbf{h}^T \cdot \mathbf{Y}_m$$

for quasi constant, **R** [(3-1) · 2]

$$\mathbf{R} = \begin{bmatrix} -1 & 1 \\ -1 & 1 \end{bmatrix}$$

and

$$\lambda = \mathrm{Tr}\,(\mathbf{h}^T \cdot \mathbf{h}) \,/\, \mathrm{Tr}\,(\mathbf{h})$$

MATLAB program gives for λ

```
>> h=[1 0;2 1;2 2]
h =
    1   0
    2   1
    2   2


>> R=[-1 1;-1 1]
R =
   -1   1
   -1   1
>> RTR=R'*R
RTR =
    2  -2
   -2   2
>> TraceRTR=2+2
TraceRTR = 4
>> lambda=14/4
lambda =
   3.5000
```

MATLAB program for lambda =3.5000

$$\mathbf{U}_{est} = (\mathbf{h}^{T} \cdot \mathbf{h} + \lambda \cdot \mathbf{R}^{T} \cdot \mathbf{R})^{-1} \cdot \mathbf{h}^{T} \cdot \mathbf{Y}_{m}$$

```
>> Uest=inv(h'*h+lambda*RTR)*h'*Ym
Uest =
  -0.2199
  -0.5183
```

MATLAB result for lambda =0 is

```
>> lambda=0
lambda = 0
>> Uest=inv(h'*h+lambda*RTR)*h'*Ym
Uest =
    2.3333
   -4.0000
>>
```

Regularization LS result for lambda=0 reduces the solution equation to

$$\mathbf{U}_{est}= (\mathbf{h}^{T} \cdot \mathbf{h})^{-1} \cdot \mathbf{h}^{T} \cdot \mathbf{Y}_{m}$$

which is the LS solution result for **U**. As expected the Regularization LS result for lambda=0 is the same as the above LS result obtained with MATLAB function h \ Y.

Regularization LS result for λ=3.5

$$\mathbf{U}_{est} = \begin{bmatrix} -0.2199 \\ -0.5183 \end{bmatrix}$$

differs from LS result due to the effect of regularization term $\lambda \cdot \mathbf{R}^{T} \cdot \mathbf{R}$ for the quasi-static regularization matrix

$$\mathbf{R} = \begin{bmatrix} -1 & 1 \\ -1 & 1 \end{bmatrix}$$

The scalar values of $\mathbf{U}_{est}$ are not the same for λ=3.5 but much close to one another then the LS values obtained for λ=0

$$\mathbf{U}_{est} = \begin{bmatrix} 2.3333 \\ -4.000 \end{bmatrix}$$

The selection of the values for $\mathbf{U}_{est}$ depend on the confidence in a posteriori the measurement data in $\mathbf{Y}_{m}$ and the a priori information used for the regularization term $\lambda \cdot \mathbf{R}^{T} \cdot \mathbf{R}$.

In the above example, measurement data in $\mathbf{Y}_m$ show significant variation

$$Ym = \begin{bmatrix} 1 \\ 2 \\ -4 \end{bmatrix}$$

and a quasi-static Regularization matrix might try to force a solution that is less realistic than the LS result.

Problems

1. Assume five measurements of the resistance versus temperature

i	1	2	3	4	5
R_i [Ω]	500	505	509	516	520
T_i [^{0}C]	29	40	49	61	70

Using MATLAB, obtain the values of the coefficients a_0 and a_1 of the linear curve $R=a_0+a_1.T$ fitting these measurements values.

2. Assume a first order instrument with impulse response

$$h(t)= \exp\{-0.1 \cdot t\}$$

and a sampling period of 2 [s].

Determine h_i for i=1, 2, ..., K, where K results from 2% settling time.

Given $U_0 = 2$, $U_1 = 4$ and $U_k = 0$ for k>1, obtain

$$Y_i = \sum_{k=i-K}^{i} h_{i-k} \cdot u_k \qquad \text{for } i = 1, 2, ..., K$$

3. Use MATLB to obtain for the following matrices

$$\mathbf{A} = \begin{bmatrix} 1 & 4 \\ 3 & 7 \end{bmatrix}$$

$$\mathbf{B} = \begin{bmatrix} 1 & 4 \\ 2 & 8 \end{bmatrix}$$

$$\mathbf{C} = \begin{bmatrix} 1 & 4 & 3 \\ 3 & 7 & 6 \end{bmatrix}$$

a) the inverse or generalized inverse
b) the norm and the rank

4. For the matrix **C** from problem 3, and

$$\mathbf{Y} = \begin{bmatrix} 2 \\ 3 \end{bmatrix}$$

obtain an estimate of U [3·1] for the underdetermined problem

$$\mathbf{Y} = \mathbf{C} \cdot \mathbf{U}$$

5. Assume, for the problem 1, the results of another set of measurements are

i	1	2	3	4	5
R_i [Ω]	500	$500+10^{-8}$	$500+10^{-9}$	$500-10^{-10}$	$500-10^{-9}$
T_i [^{0}C]	29	40	49	61	70

Using MATLAB, obtain the values of the coefficients a_0 and a_1 of the linear curve $R=a_0+a_1.T$ fitting these measurements values.

6. Given

$$\mathbf{h} = \begin{bmatrix} 2 & 1 \\ 3 & 0 \\ 3 & 7 \end{bmatrix}$$

and using MATLAB, obtain

a) The generalized inverse using SVD method.

b) $\mathbf{U_{est}} = \mathbf{h^{-g}} \cdot \mathbf{Y_m}$ for

$$\mathbf{Y}_m = \begin{bmatrix} 2.00 \\ 2.10 \\ 1.99 \end{bmatrix}$$

b) The regularization solution

$$\mathbf{U_{est}} = (\mathbf{h}^T \cdot \mathbf{h} + \lambda \cdot \mathbf{R}^T \cdot \mathbf{R})^{-1} \cdot \mathbf{h}^T \cdot \mathbf{Y_m}$$

for

$$\mathbf{R} = \begin{bmatrix} -1 & 1 \\ -1 & 1 \end{bmatrix}$$

Chapter 4

Inverse Problems in Dynamic Calibration of Sensors

4.1 Introduction

System monitoring requires numerous sensors but rarely is possible to have as many sensors as quantities to measure. Frequently, system monitoring problem is under-determined. Moreover, time varying signals require dynamic measurement, while sensors are characterized by bandwidth frequencies that can be lower than the useful range of frequencies of the signals to be measured. Computer based instrumentation can alleviate such difficulties by providing means for estimating system variables that are not directly measured, *i.e.* indirect measurement, and by increasing the range of signal frequencies the sensors can measure accurately. In this section these problems are analyzed using ill-posed inverse problems theory in particular using inverse dynamics approaches. This chapter presents in the beginning the case of simple first and second order instruments considered alone, then continues with the investigation of the full order and reduced order dynamic calibration (compensator) for sensors. In the first part of this chapter, noiseless sensing and exact models are assumed in order to focus the analysis of inverse problem issues for first and second order instruments [71]. For this purpose, complex domain and frequency domain analysis are carried out for lumped parameters models of sensors. Section 4.2 presents the analysis of first and second order instruments dynamics simulation using transfer functions, time response and Bode diagrams. In section 4.3 the investigation is extended by including dynamic calibration, anti-aliasing filters and phase lead compensators required in computer based instrumentation. In section 4.4, the effect of measurement noise on dynamic calibration is analyzed. In section 4.5, the focus is on state estimation in indirect sensing.

Forward Dynamics of a sensor, modeled as a LTI system, can be described by its transfer function G(s) = Y(s) / X(s), with X(s) = input and Y(s) = output. First and second order instruments analysis will illustrate general issue in dynamic calibration using inverse models.

4.2 First Order Instruments

4.2.1 *Time and Frequency Response of Forward Dynamics*

Numerous sensors can be modeled by a first order transfer function

$$G(s) = \frac{k}{1 + T \cdot s}$$

where k = gain, T = time constant.

Example 4.1 The first order model for a J thermocouple with gain at 20 [^{0}C] of =50 [μV/^{0}C] and time constant of 0.01 [s] is assumed, i.e. k = 5 [10 μV/^{0}C] and T = 0.01 [8].

MATLAB program is

```
k =5;
T=0.01;
num=[0 k];
den=[T 1];
step(num,den);grid
```

Figure 4.1 shows the Y(t) plot of the unit step response of a first order instrument.

MATLAB program

```
>> bode(num,den);
grid;
```

The Bode diagram results for the above case is shown in Fig. 4.2.

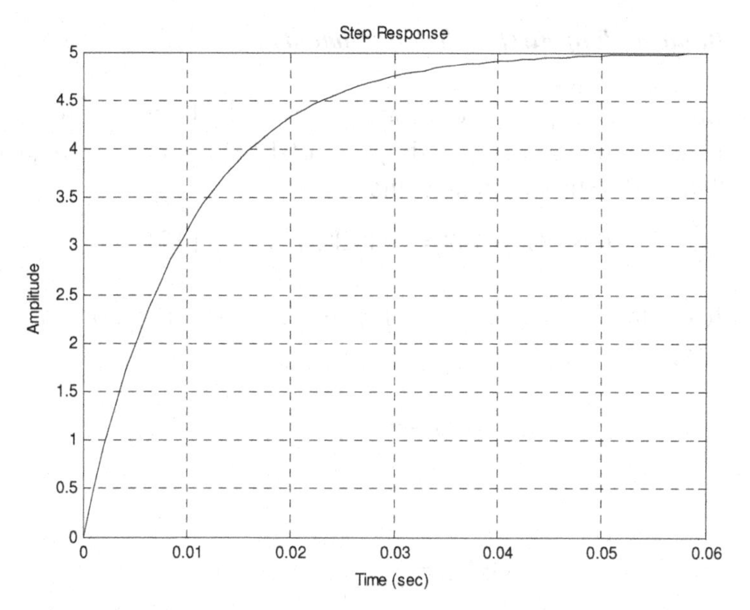

Fig. 4.1 The plot of the unit step response Y(t) of a first order instrument

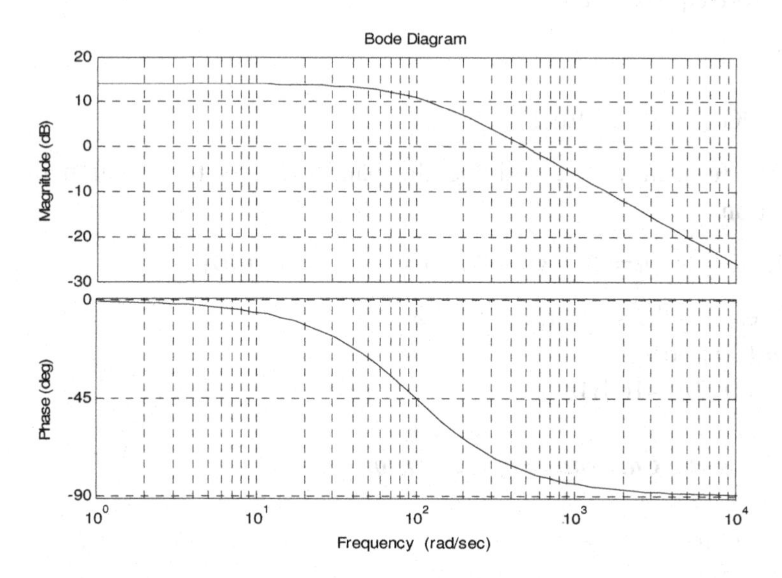

Fig. 4.2 Bode diagram for the case shown in Fig. 4.1

4.2.2 *Bandwidth of First Order Instruments*

Bandwidth cutoff frequency, ω_b, corresponds to the frequency for which the Magnitude $20 \log|G(j \cdot \omega)|$ drops 3 [dB] below its zero-frequency value $20 \log|G(j \cdot 0)|$, such that, at the limit,

$$20 \log|G(j \cdot \omega_b)| = 20 \log|G(j \cdot 0)| - 3 \text{ [dB]}$$

Example 4.2 In the above case of a first order transfer function for k = 5, T = 0.01 [s]

$$G(s) = \frac{5}{1 + 0.01 \cdot s}$$

or

$$G(j\omega) = \frac{5}{1 + 0.01 \cdot j \cdot \omega}$$

its zero-frequency value is

$G(j0) = 5$

or

$20 \log|G(j \cdot 0)| = 14$ [dB]

Cutoff frequency, ω_b, defining the bandwidth, is the solution of the equation

$20 \log|G(j \cdot \omega_b)| = 20 \log|G(j0)| - 3 = 14 - 3 = 11$ [dB]

The result is the same as in Fig. 4.2.

$\omega_b = 100$ [rad/s]

$f_b = 100/(2\pi) \approx 16$ [Hz]

4.2.3 *Static Calibration of the Sensor*

Assume that

X_n is unknown sensor input to the sensor

Y_m is the measured output from the sensor.

Static calibration uses steady state sensor response to unit step, $X(s) = 1 / s$, for determining the value of K_c for calibration, *i.e.* for estimating the unknown input, using final value theorem

$$K_c = \lim_{s\to\infty} Y(t) = \lim_{s\to 0} s \cdot G(s) \cdot X(s) = \lim_{s\to 0} s \cdot G(s) \cdot \frac{1}{s} = G(0)$$

such that estimated input value using static calibration is given by

$$X(t) = \frac{Y(t)}{K_c} = \frac{L^{-1} Y(s)}{K_c}$$

where L^{-1} is inverse Laplace transform operator.

Example 4.3 Estimate input value using static calibration.

Estimated input value using static calibration, for the above Example 4.2, for static calibration constant

$$K_c = G(j \cdot 0) = 5$$

gives the estimated input value

$$X_{est}(t) = \frac{Y(t)}{5}$$

Consequently, the result for Y(t) shown in Fig. 4.1 can be converted in $X_{est}(t)$ by dividing the Amplitude scale by 5, to obtain a scale for $X_{est}(t)$ that reaches steady state value of 1.

This $X_{est}(t)$ is an estimation of the unit step input that has zero estimation error only at steady state. The result shows that $X_{est}(t)$ becomes accurate only after settling time, for example with an error of 2% after four time constants, $4 \cdot T = 4 \cdot 0.01 = 0.04$ [s].

In practice, for sinusoidal inputs, the estimation error is considered acceptable within bandwidth, i.e. for input signal frequencies

$$\omega < \omega_b$$

where, in this case

$$\omega_b = 100 \text{ [rad/s]}$$

This issue is analyzed in the next paragraph.

4.2.4 *Sinusoidal Response of the Sensor-MATLAB Simulations*

The block diagram of a first order instrument with sinusoidal input is shown in Fig. 4.3.

Given that the Laplace transform $X(s) = L\{X(t)\}$ for $X(t) = \sin \omega t$ is

$$X(s) = \omega / (s^2 + \omega^2)$$

and for unit impulse input

$$\delta(s) = 1$$

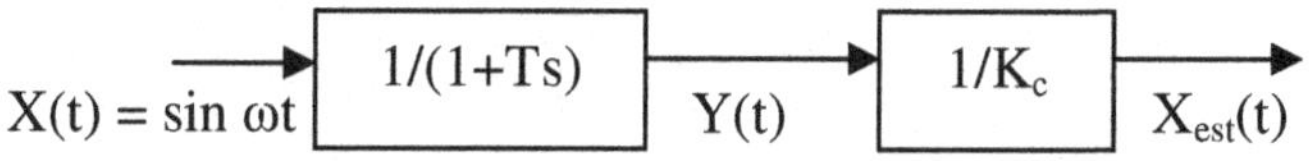

$$X(t) = \sin \omega t \rightarrow \boxed{1/(1+Ts)} \xrightarrow{Y(t)} \boxed{1/K_c} \rightarrow X_{est}(t)$$

Fig. 4.3 Block diagram of a first order instrument with sinusoidal input

MATLAB simulation uses the computation scheme from Fig. 4.4.

In fact

$$L^{-1}\{\delta(s) \cdot \omega / (s^2 + \omega^2)\} = L^{-1}\{\omega / (s^2 + \omega^2)\} = \sin \omega t$$

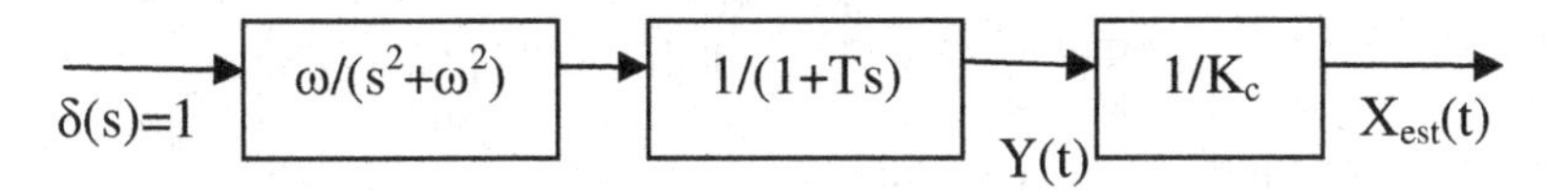

Fig. 4.4 MATLAB simulation scheme

Example 4.4 MATLAB Simulation for k = 5; T = 0.01; K_c = 5 and various values of ω.

1) ω = 10 [rad/s]
   MATLAB program is

```
num=[0 0 0 50];
den=[0.01 1 1 100];
impulse(num,den);
```

Figure 4.5 shows the plot Y(t) for s sinusoidal input with ω = 10 [rad/s].

2) ω = 100 [rad/s]
   MATLAB program is

```
num=[0 0 0 500];
den=[0.01 1 100 10000];
impulse(num,den);
```

Figure 4.6 shows the plot Y(t) for s sinusoidal input with ω = 100 [rad/s].

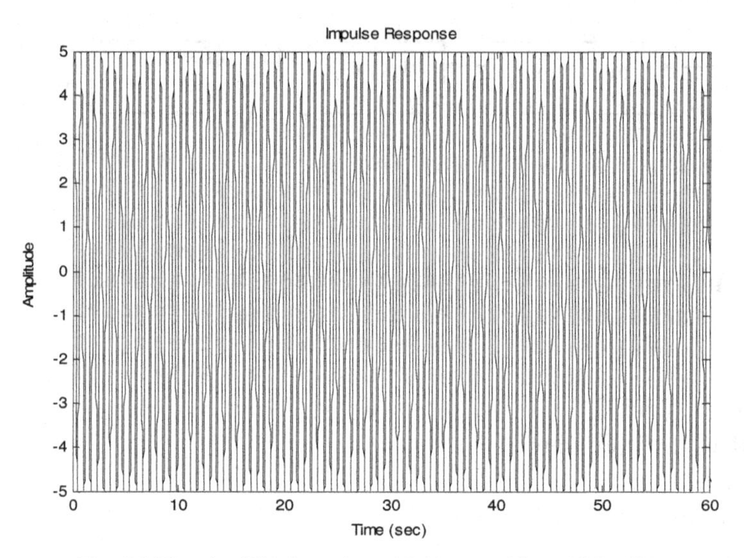

Fig. 4.5 The plot Y(t) for s sinusoidal input with ω=10 [rad/s]

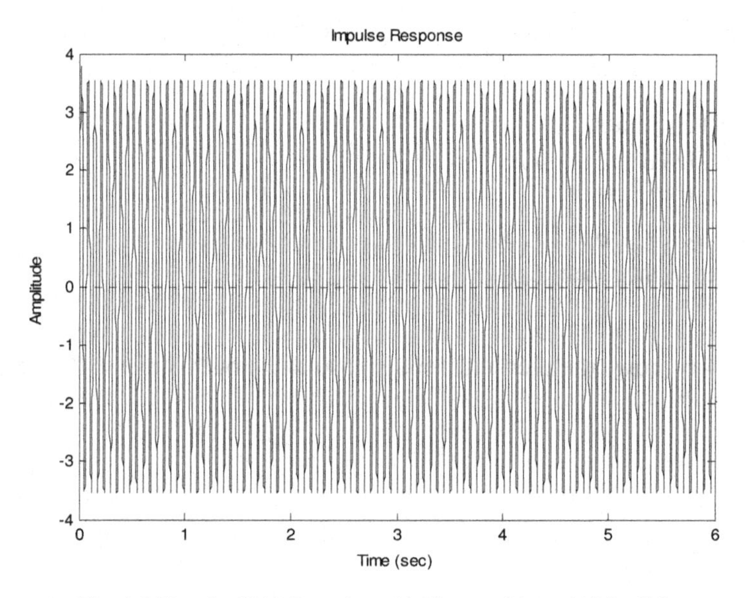

Fig. 4.6 The plot Y(t) for s sinusoidal input with ω=100 [rad/s]

3) $\omega = 1000$ [rad/s]

MATLAB program is

```
num=[0 0 0 5000];
den=[0.01 1 10000 1000000];
impulse(num,den);
```

Figure 4.7 shows the plot Y(t) for s sinusoidal input with $\omega = 1000$ [rad/s].

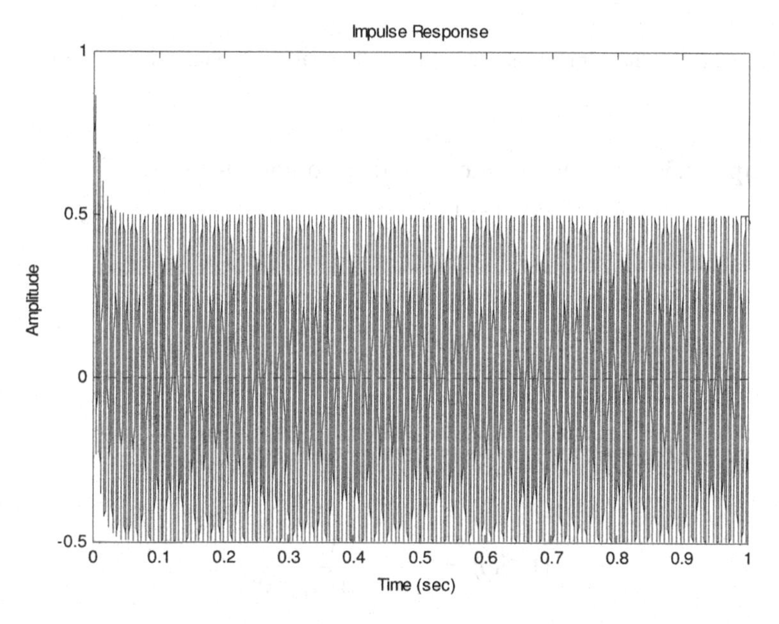

Fig. 4.7 The plot Y(t) for s sinusoidal input with ω=1000 [rad/s]

These results illustrate the significant decrease of the amplitude $|X_{ext}(t)|$ as ω increases beyond $\omega_b = 100$ [rad/s].

4.2.5 *Analytical Solutions for Harmonic Response of First Order Instruments*

The block diagram for the derivation of the analytical solution for the harmonic response of a first order system is shown in Fig. 4.8.

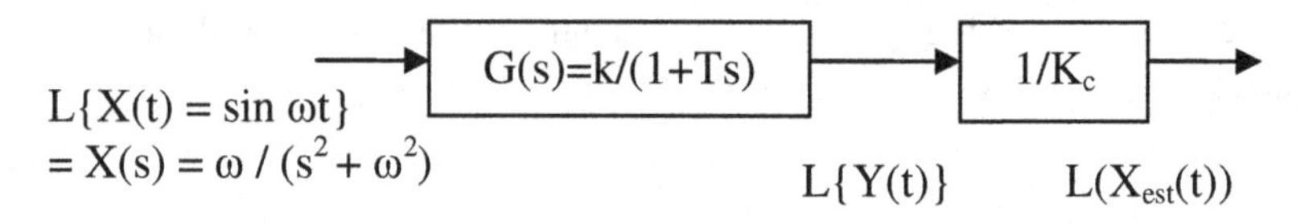

Fig. 4.8 Block diagram for the derivation of the analytical solution for the harmonic response

Example 4.5 Derivation of the analytical solution for $k = 5$, $T = 0.01$ and $X(t) = 1 \cdot \sin \omega \cdot t$, *i.e.* $|X(t)| = 1$

For

$$G(s) = \frac{k}{1 - T \cdot s}$$

$$G(j\omega) = \frac{k}{1 + T \cdot j\omega} = \frac{5}{1 + 0.01 \cdot j\omega} = \frac{5 \cdot (1 - j \cdot 0.01 \cdot \omega)}{1 + 0.01^2 \cdot \omega^2}$$

$$|G(j\omega)| = \frac{5}{(1 + 0.01^2 \cdot \omega^2)^{1/2}}$$

$$\Phi = \tan^{-1}(-0.01\omega)$$

static calibration uses
$K_c = G(j0) = k = 5$

The output $Y(t)$ is given by [50]
$y(t) = 1 \cdot |G(j\omega)| \cdot \sin(\omega \cdot t + \Phi)$

while estimated input is

$$X_{est}(t) = \frac{y(t)}{K_c} = \frac{|G(j\alpha)|}{K_c}\sin(\omega \cdot t + \Phi)$$
$$= \frac{|G(j\omega)|}{5}\sin(\omega \cdot t + \tan^{-1}(0.01 \cdot \omega))$$

such that

$$\frac{|X_{est}(t)|}{|X(t)|} = \frac{|G(j\omega)|}{5}$$

For $\omega = 10$ the result is

$$|G(j\omega)| = 5/(1 + 0.01^2\omega^2)^{1/2} = 5/(1 + 0.01^2 10^2)^{1/2} \approx 5$$

$$\Phi = \tan^{-1}(- 0.01 \cdot \omega) = - 5.71$$

$$y(t) = 1\ |G(j\omega)| \sin(\omega \cdot t + \Phi) \approx 5 \sin \omega \cdot t$$

$$x_{est}(t) = y(t) / k \approx 1 \sin \omega \cdot t = x(t)$$

$$|x_{est}(t)| / |x(t)| \approx 1$$

For $\omega = 100$, the result is

$$|x_{est}(t)| / |x(t)| \approx 0.7$$

$$20 \log |x_{est}(t)| / |x(t)| = - 03\ [\text{dB}], \quad \text{i.e. } \omega = 100 = \omega_b$$

Summary of results for $\omega = 10$, 100 and 1000 [rad/s] is the following

ω [rad/s]	f[Hz]	1/f[s]	$\lvert G(j\omega)\rvert$	Φ	y(t)	$x_{est}(t)$	$\lvert x_{est}(t)\rvert/\lvert x(t)\rvert$
10	1.58	0.63	≈5	≈ 0	≈5sin ωt	≈sin ωt	≈1
100	15.8	0.063	≈3.5	≈-45	3.5sin(ωt-45)	0.7sin(ωt-45)	≈0.7
1000	158	0.0063	≈3.5	≈-90	0.5sin(ωt-90)	0.1sin(ωt-45)	≈0.1

Consequently, for

$$\begin{array}{ll} \omega > \omega_b & |x_{est}(t)| / |x(t)| < 0.7 \\ \omega >> \omega_b & |x_{est}(t)| / |x(t)| << 0.7 \\ \omega = 1000 & |x_{est}(t)| / |x(t)| \approx 0.1 \end{array}$$

i.e. for $\omega = 1000$ the estimated $|x_{est}(t)|$ is only 10% of the amplitude of the sensor input signal $|x(t)|$.

Dynamic estimation (calibration) can be achieved using the inverse problem solution. It can be observed that dynamic calibration results in increasing gains, in this case $1 / 0.7 = 1.43$ for $\omega = 100$ and $1 / 0.1 = 10$ for $\omega = 1000$ [71].

Obviously, these gains increase with ω, and this can lead to various difficulties (overflow in numerical computations, over-amplification of noise high frequency-low amplitude components in the y(t) signal etc), to be addressed by the solutions to ill-posed problems from Ch. 3.

4.3 Second Order Instruments

4.3.1 *Static Calibration*

Second order transfer function [50], for a mass-spring-damper system, is

$$G(s) = \frac{k}{s^2 + 2 \cdot \varsigma \cdot \omega_n \cdot s + \omega_n^2}$$

where k = gain, ω_n = un-damped natural frequency, ζ = damping ratio.

An example could be a force f(t) transducer [18], with the block diagram shown in Fig. 4.9 where
f(t) [N] is the input force to measure
d(t) [m] is the output displacement
v(t) [V] is the output voltage of the potentiometer.

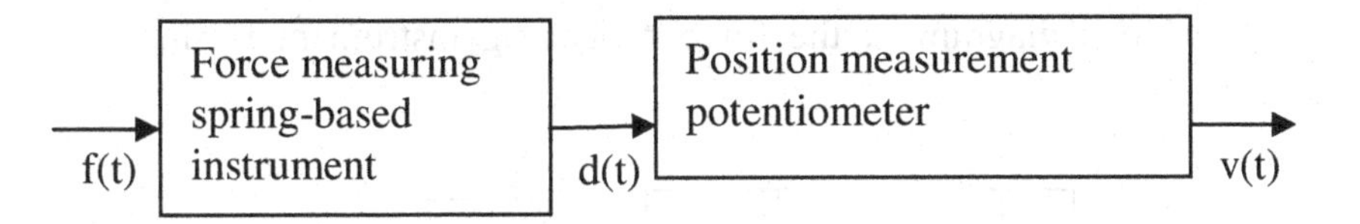

Fig. 4.9 Block diagram of a force transducer

Assume that the force measuring spring based instrument measured is modeled by a mass-spring-damper M-B-K system in horizontal motion (such that gravity effect can be ignored in deriving motion equation). The transfer function is

$$\frac{f(s)}{d(s)} = \frac{1}{M \cdot s^2 + B \cdot s + K}$$

where

$$d(s) = L\{d(t)\}$$
$$f(s) = L\{f(t)\}$$

Assume the approximate transfer function of the position measurement potentiometer

$$v(s)/d(s) = K_p$$

where

$$v(s) = L\{v(t)\}$$
$$d(s) = L\{d(t)\}$$

where K_p [V/m] is the calibration constant of the potentiometer.

The block diagram of the force measuring instrument is shown in Fig. 4.10.

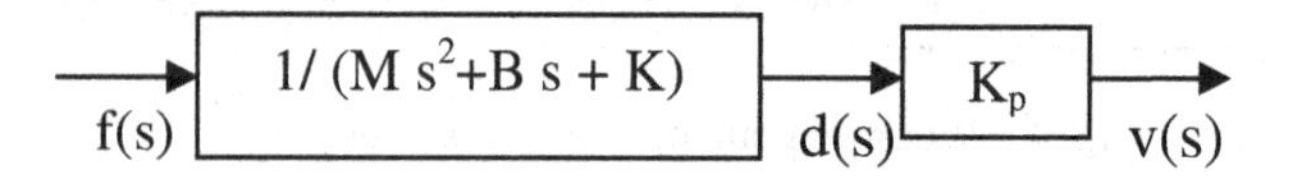

Fig. 4.10 Block diagram of the force measuring instrument

The overall transfer function is

$$G(s) = \frac{v(s)}{f(s)} = \frac{K_p}{M \cdot s^2 + B \cdot s + K}$$

$$= \frac{\frac{K_p}{M}}{s^2 + \frac{B}{M} \cdot s + \frac{K}{M}} = \frac{k}{s^2 + 2 \cdot \varsigma \cdot \omega_n \cdot s + \omega_n^{\ 2}}$$

where

$$\omega_n^{\ 2} = K / M$$
$$2 \cdot \zeta / \omega_n = B / M$$
$$k = K_p / M$$

Time response of such second order instruments is strongly dependent on the value of

$$\varsigma = \frac{B \cdot \omega_n}{2 \cdot M}$$

The damping ratio, ζ, determines the type of response:

$\zeta < 1$ under-damped response
$\zeta = 1$ critically damped response
$\zeta > 1$ over-damped response

Example 4.6 MATLAB Simulations for frequency response of sensors forward dynamics and compensators based on inverse dynamics for a second order instrument with k = 1, ω_n = 10 [rad/s] f_n = 1.58 [Hz] period of 0.63 [s] and various values of the damping ratio ζ.

Second order transfer function is

$$G(s) = \frac{v(s)}{d(s)} = \frac{k}{s^2 + b \cdot s + c} = \frac{k}{s^2 + 2 \cdot \varsigma \cdot \omega_n \cdot s + \omega_n^{\ 2}}$$

where

$$b = 2 \cdot \zeta \cdot \omega_n$$
$$c = \omega_n^{\ 2}$$

Steady state value of v(t) for unit step input f(t) is obtained using limit value theorem for unit step input f(s) = 1 / s for s tending towards zero

$$v_{ss} = \lim_{s \to 0} s \cdot G(s) \cdot \frac{1}{s} = \lim_{s \to 0} \frac{k}{s^2 + 2 \cdot \varsigma \cdot \omega_n \cdot s + \omega_n^{\ 2}} = \frac{1}{\omega_n^{\ 2}} = 0.01$$

The computation is carried out for ζ = 0, 0.1, 0.6, 1.2.

a) for ζ = 0, b = 2 · ζ · ω_n = 0
MATLAB program is
```
k=1;
b=0;
c=100;
num=[ 0 0 k];
den=[1 b c];
step(num,den);grid
```

Figure 4.11 shows the plot of the un-damped oscillatory response v(t), obviously not useful in practical applications. Higher values for ζ are required.

b) for $\zeta = 0.1$, $b = 2 \cdot \zeta \cdot \omega_n = 2$

MATLAB program is

```
k=1;
b=2;
c=100;
num=[ 0 0 k];
den=[1 b c];
step(num,den);grid
```

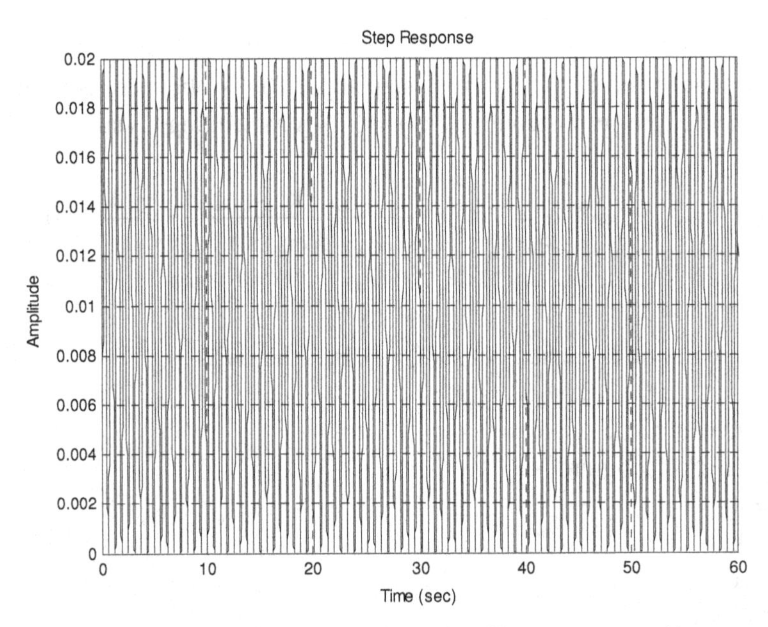

Fig. 4.11 The plot of the un-damped oscillatory response v(t)

The results from Fig. 4.12 show significant maximum overshoot of 70% and long 2% settling time of

$4/(\zeta\,\omega_n) = 4/\,(0.1 \cdot 10) = 4$ [s].

c) for $\zeta = 0.6$, $b = 2 \cdot \zeta \cdot \omega_n = 12$
MATLAB program is

```
k=1;
b=12;
c=100;
num=[ 0 0 k];
den=[1 b c];
step(num,den);grid
```

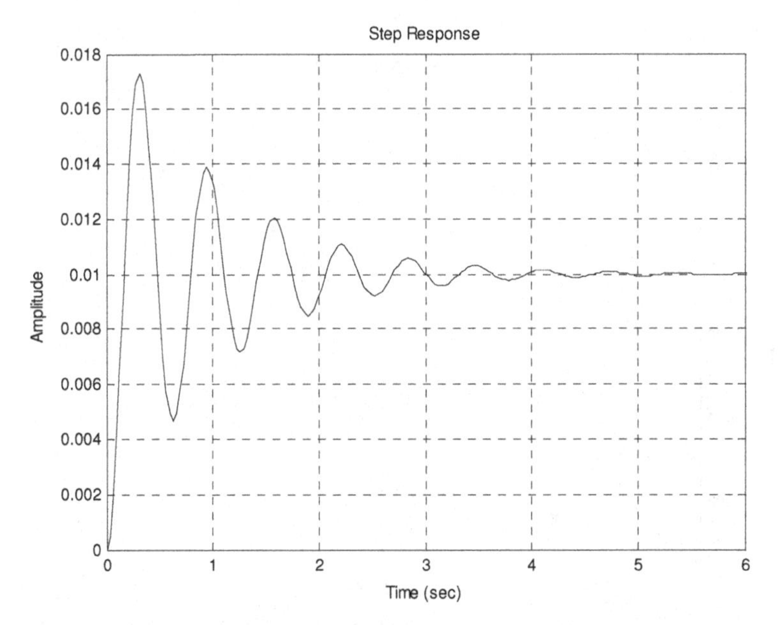

Fig. 4.12 The plot of the oscillatory response v(t) for $\zeta = 0.1$

The results from Fig. 4.13 show significant reduced maximum overshoot of 5% and reduced 2% settling time to $4\,/\,(\zeta \cdot \omega_n) = 4\,/\,(0.6 \cdot 10) = 0.67$ [s].

d) for over-damped case, $\zeta = 1.2$, $b = 2 \cdot \zeta \cdot \omega_n = 24$
MATLAB program is

```
k=1;
b=24;
c=100;
num=[ 0 0 k];
den=[1 b c];
step(num,den);grid
```

The results from Fig. 4.14 show no overshoot but sluggish response.

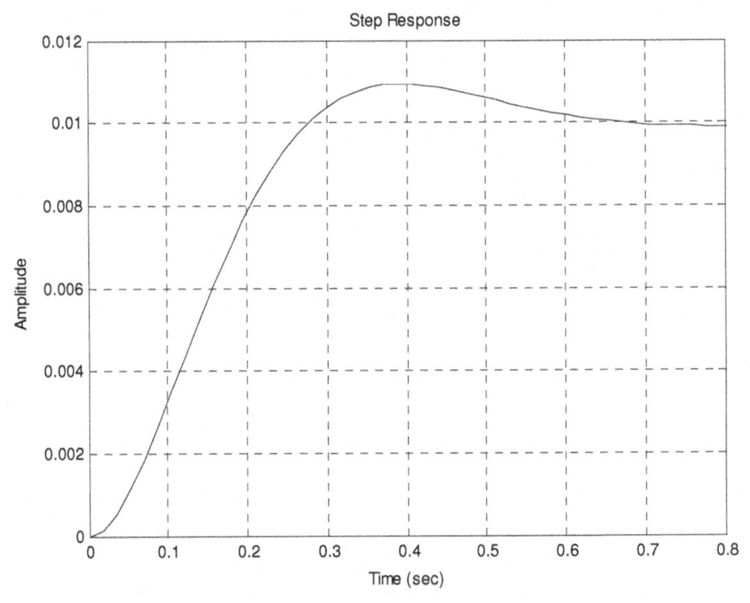

Fig. 4.13 The plot of the oscillatory response v(t) for $\zeta = 0.6$

Present day practice is to provide under-damped response for $\zeta = B \cdot \omega_n / (2 \cdot M)$ in the range of 0.6-0.7, by selecting a damping coefficient of $B = 2 \cdot M \cdot \zeta / \omega_n$ [71].

Example 4.7 Bode diagram for $\zeta = 0.6$, b= $2 \cdot \zeta \cdot \omega_n = 12$ is given by the MATLAB program:

```
k=1;
b=12;
c=100;
num=[ 0 0 k];
den=[1 b c];
bode(num,den);
grid;
```

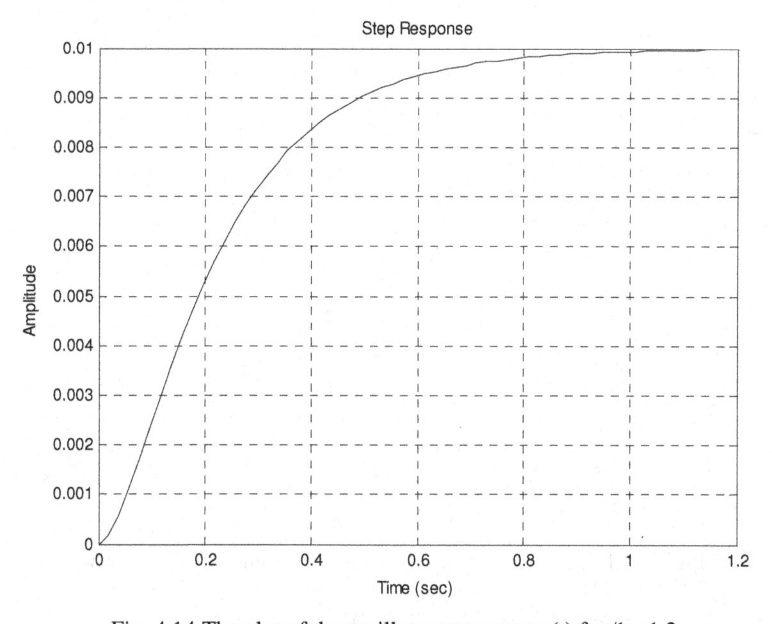

Fig. 4.14 The plot of the oscillatory response v(t) for $\zeta = 1.2$

Bode diagram of a second order instrument with $\zeta = 0.6$ is shown in Fig. 4.15.

Magnitude response is flat up to approx. 10 [rad/s], while the phase lag becomes noticeable after 1 [rad/s].

4.3.2 *Harmonic Response of the Second Order Sensor with ζ =0.6. MATLAB Simulations*

Time response of such second order instruments depends significantly on the value of ζ.

MATLAB simulation is carried out for the second order instrument subject to static calibration, shown in Fig. 4.16.

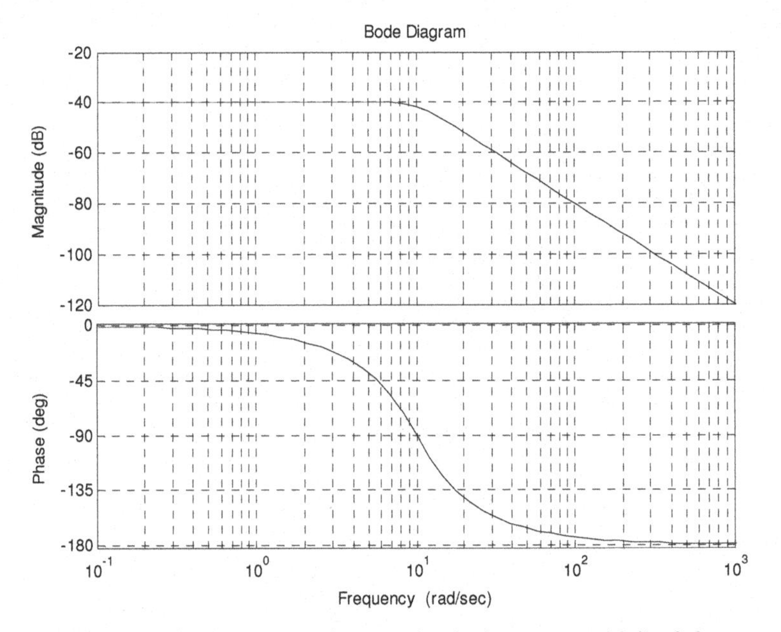

Fig. 4.15 Bode diagram of a second order instrument with $\zeta = 0.6$

Static calibration constant is $K_c = G(j \cdot 0)$.

For $X(s) = \omega / (s^2 + \omega^2)$ and unit impulse input $\delta(s) = 1$, the harmonic response simulation is achieved with MATLAB instruction impulse (num, den) for the functions shown in Fig. 4.17.

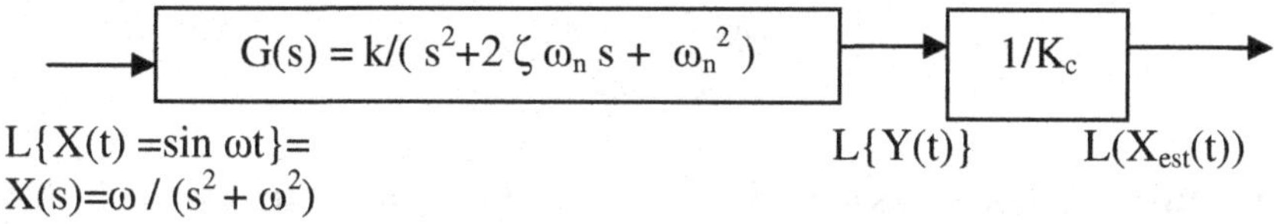

$G(s) = k/(s^2+2\,\zeta\,\omega_n\, s + \omega_n^2)$ → $1/K_c$ →

$L\{X(t) = \sin \omega t\} = X(s) = \omega / (s^2 + \omega^2)$ $L\{Y(t)\}$ $L(X_{est}(t))$

Fig. 4.16 Second order instrument subject to static calibration

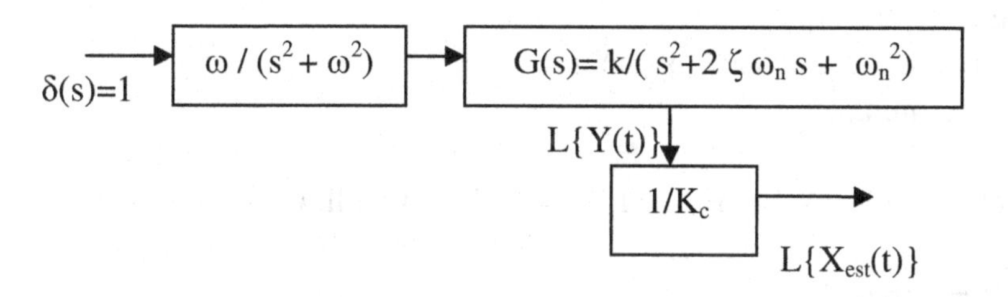

Fig. 4.17 MATLAB simulation model

The overall transfer function $X_{est}(s) / \delta(s)$ for the MATLAB program is

$$\frac{X_{est}(s)}{\delta(s)} = \frac{k \cdot \omega}{(s^2 + \omega^2) \cdot (s^2 + 2 \cdot \varsigma \cdot \omega_n \cdot s + \omega_n{}^2)}$$

$$= \frac{k \cdot \omega}{s^4 + 2 \cdot \varsigma \cdot \omega_n \cdot s^3 + (\omega_n^2 + \omega^2) \cdot s^2 + 2 \cdot \varsigma \cdot \omega_n \cdot \omega^2 \cdot s + \omega_n{}^2 \cdot \omega^2}$$

or

$$\frac{X_{est}(s)}{\delta(s)} = \frac{k \cdot \omega}{s^4 + a \cdot s^3 + (b + \omega^2) \cdot s^2 + a \cdot \omega^2 \cdot s + b \cdot \omega^2}$$

where

$$a = 2 \cdot \zeta \cdot \omega_n$$
$$b = \omega_n^2$$

Example 4.8 Let us assume k = 1, $\zeta = 0.6$, $\omega_n = 10$ [rad/s] such that $a = 2 \cdot \zeta \cdot \omega_n = 12$ and $b = \omega_n^2 = 100$.

In MATLAB notation

```
num=[ 0 0 0 0 ω];
den=[1 a b+ ω^2 a* ω^2  b* ω^2];
impulse(num,den);
```

MATLAB results for ω=1 and 5 [rad/s] are the following

1) ω = 1 [rad/s]

   MATLAB program is

```
a=12;
b=100;
num=[ 0 0 0 0 1];
den=[1 a b+1 a  b];
impulse(num,den);
```

   The plot is shown in

The amplitude from Fig. 4.18 agree to amplitude $|G(j \cdot \omega|$ for Bode diagram from Fig. 4.15 for ω = 1 that is -40 [dB]= 20 log 0.01.

2) ω = 5 [rad/s]

```
a=12;
b=100;
num=[ 0 0 0 0 5];
den=[1 12 125 300 2500];
impulse(num,den);
```

The amplitude from Fig. 4.19 agrees again to amplitude $|G(j \cdot \omega|$ for Bode diagram from Fig. 4.15 for ω = 5 that is -40 [dB] = 20 log 0.01.

4.3.3 *Analytical Solutions for Harmonic Response and Bandwidth Frequency of a Second Order Instrument*

The second order instrument transfer function is

$$G(s) = \frac{k}{s^2 + 2 \cdot \varsigma \cdot \omega_n \cdot s + \omega_n^{\ 2}}$$

Examples of analytical calculation of the harmonic response and bandwidth frequency are presented in the following Example 4.9.

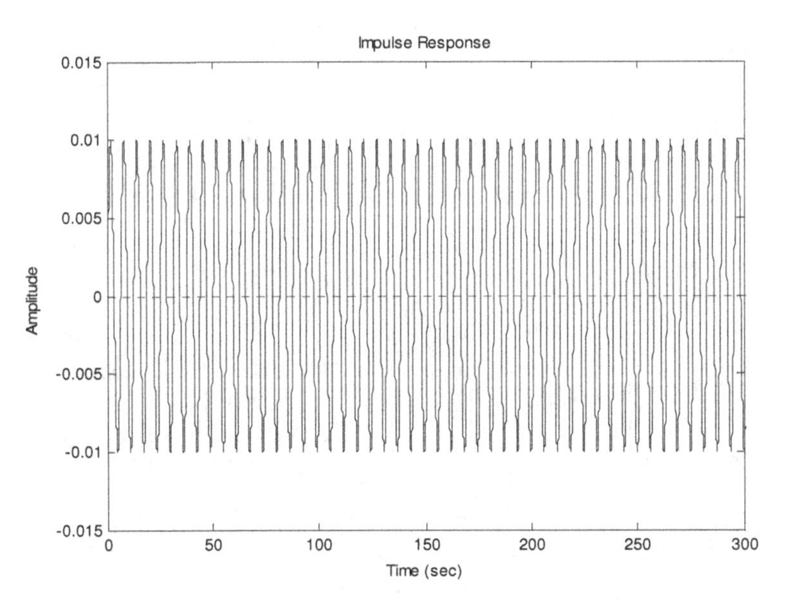

Fig. 4.18 The time response for ω=1 rad/s

Example 4.9 Analytical results for a second order instrument.

a) Harmonic response of a second order system for various values of ω

Consider the block diagram from Fig. 4.16 with

$X(t) = 1 \cdot \sin \omega t$

$|X(t)| = 1$

$k = 1$

$\zeta = 0.6$

$\omega_n = 10$ [rad / s]

$2 \cdot \zeta \cdot \omega_n = 12$

$\omega_n^2 = 100$

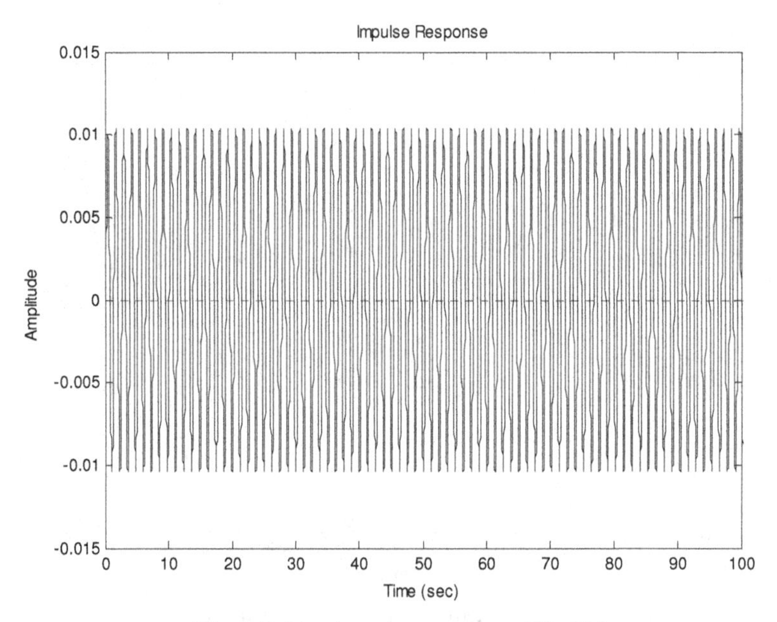

Fig. 4.19 The time response for ω=5 [rad/s]

The transfer function is

$$G(s) = \frac{1}{s^2 - 12 \cdot s - 100}$$

$$G(j\cdot\omega) = \frac{1}{-\omega^2 + 100 + 12\cdot j\cdot\omega}$$

$$= \frac{-\omega^2 + 100 - 12\cdot j\cdot\omega}{(-\omega^2 + 100 + 12\cdot j\cdot\omega)(-\omega^2 + 100 - 12\cdot j\cdot\omega)}$$

$$= \frac{-\omega^2 + 100 - 12\cdot j\cdot\omega}{[(-\omega^2 + 100)^2 + 12^2\cdot\omega^2]}$$

$$|G(j\cdot\omega)| = \frac{1}{[(-\omega^2 + 100)^2 + 12^2\cdot\omega^2]^{1/2}}$$

$$|G(j0)| = 1/((100)^2)^{1/2} = 1/100$$

$$K_c = G(j0) = 1/100$$

$$\Phi = \tan^{-1}\left(\frac{-12\cdot j\cdot\omega}{-\omega^2 - 100}\right)$$

Y(t) is given by [50]

$$Y(t) = 1\cdot|G(j\omega)|\cdot\sin(\omega\cdot t + \Phi)$$

while estimated input is given by static calibration as

$$X_{est}(t) = \frac{Y(t)}{K_c} = \frac{|G(j\cdot\omega)|)}{K_c}\sin(\omega\cdot t + \Phi) = \frac{|G(j\cdot\omega)|)}{1/100}\sin(\omega\cdot t + \Phi)$$

such that

$$|X_{est}(t)| = \frac{|G(j\cdot\omega)|}{K_c} = \frac{|G(j\cdot\omega)|}{|G(j\cdot 0)|} = \frac{|G(j\cdot\omega)|}{1/100}$$

b) Harmonic response for $\omega = 10$ is given by

$$|G(j\cdot\omega)| = \frac{1}{[(-\omega^2+100)^2+12^2\cdot\omega^2]^{1/2}} = \frac{1}{[(-10^2+100)^2+12^2\cdot 10^2]^{1/2}} = \frac{1}{120}$$

$$\Phi = \tan^{-1}\left(\frac{-12\cdot j\cdot\omega}{-\omega^2-100}\right) = \tan^{-1}\left(\frac{-j\cdot 12\cdot 10}{-10^2-100}\right) = \tan^{-1}\left(\frac{-j\cdot 12\cdot 10}{0}\right) = -90^0$$

and

$$20\log|G(j\omega)| = 20\log(1/120) = 41.6\ [\text{dB}]$$

For $\omega = 10$, Bode diagram in Fig. 4.15 gives approx -42 [dB] and -90^0 which agrees with the above simulation results.

$$Y(t) = 1\,|G(j\omega)|\cdot\sin(\omega t+\Phi) = ((1/100))\cdot\sin(\omega t-90)$$

$$|X_{est}(t)| = \frac{|G(j\cdot\omega)|}{K_c} = \frac{|G(j\cdot 10)|}{|G(j\cdot 0)|} = \frac{1/120}{1/100} = 0.833$$

For, $|X(t)|=1$

$$\frac{|X_{est}(t)|}{|X(t)|} = 0.833$$

c) Calculation of the cutoff frequency ω_b

Exact calculation of the cutoff frequency, ω_b, defining the bandwidth, is obtained from the equation of definition of bandwidth [50]

$$20\cdot\log|G(j\cdot\omega_b)| = 20\cdot\log|G(j\cdot 0)| - 3$$

or

$$\frac{|G(j \cdot 0)|}{|G(j \cdot \omega_b)|} = \log^{-1} \frac{3}{20} = 1.4125$$

or

$$|G(j \cdot \omega_b)| = 0.709 \cdot |G(j \cdot 0)|$$

which shows that, at cutoff frequency, ω_b, the amplitude $|G(j \cdot \omega_b)|$ drops to 0. 709 of the $|G(j \cdot 0)|$

For the above second order instrument

$$|G(j \cdot \omega)| = \frac{1}{[(-\omega^2 + 100)^2 + 12^2 \cdot \omega^2]^{1/2}}$$

and

$$|G(j \cdot 0)| = \frac{1}{[(-0^2 + 100)^2 + 12^2 \cdot 0^2]^{1/2}} = \frac{1}{100}$$

The cutoff frequency, ω_b can be obtained from

$$|G(j \cdot \omega_b)| = \frac{1}{[(-\omega_b{}^2 + 100)^2 + 12^2 \cdot \omega_b{}^2]^{1/2}} = 0.709 \cdot |G(j \cdot 0)| = \frac{0.709}{100}$$

The solution ω_b of the equation is obtained as follows

$$[(-\omega_b{}^2 + 100)^2 + 12^2 \cdot \omega_b{}^2]^{1/2} = \frac{100}{0.709}$$

$$\omega_b{}^4 - 56\, \omega_b{}^2 - 9952.6 = 0$$

The solution for $\omega_b{}^2$ is

$$\omega_b{}^2 = 28 \pm \sqrt{(28^2 + 9952.6)} = 28 \pm 103.6 \quad \text{or } 131.6 \text{ and } -75.6$$

Only the positive real solution is retained

$$\omega_b = +\sqrt{131.6} = 11.47 \text{ [rad / s]}$$

This result is the same as from the second order Bode diagram from Fig. 4.15 for the magnitude of

$$20 \cdot \log |G(j \cdot \omega_b)| = 20 \cdot \log |G(j \cdot 0)| - 3 = -40 - 3 = -43 \text{ [dB]}.$$

4.4 Calibration for Computer-Based Instrumentation

In this section, for making possible to present the main features of static and dynamic calibration approaches, only linear time invariant detector models for each pixel of the image, limited to given operational frequency domains are considered, in order to achieve constant gain over this frequency domain and the same phase shift.

Association and fusion of these signals require first signal processing of different discrete spatial time representations that are specific to various instruments, such that these variables, measured by multiple sensors, will be referenced to the same spatial coordinates and will be synchronized in time [61]. Sensors outputs are dependent, however, not only on the inputs from measured target variables, but also on the instrument design and sensor dynamics. Sensor fusion is accurate only if it uses signals that are properly calibrated and compensated for the phase difference [65]. An effective approach for achieving these requirements is dynamic calibration of individual sensors output signals.

Dynamic calibration is investigated as an inverse problem which permits to use numerical solutions already developed for such problems. Numerical results illustrate the benefits of dynamic calibration for various sensors. Dynamic calibration is proposed in order to improve the measurement accuracy, compensation for phase lag for sensor fusion and phase etc. (Fig. 4.20 and Fig. 4.21).

The focus in the chapter is on the investigation of the linear time invariant (LTI) sensor models for dynamic calibration (See Fig. 4.21).

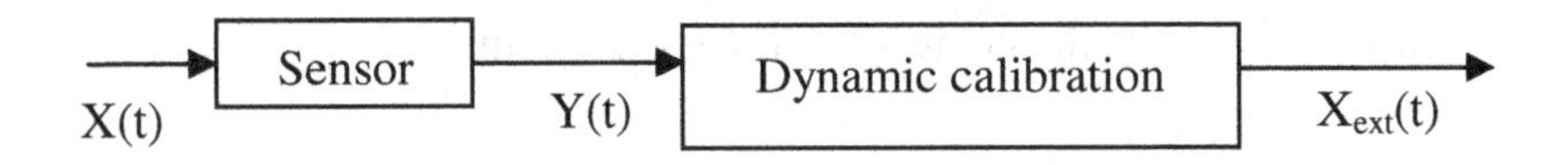

Fig. 4.20 Block diagram of dynamic calibration

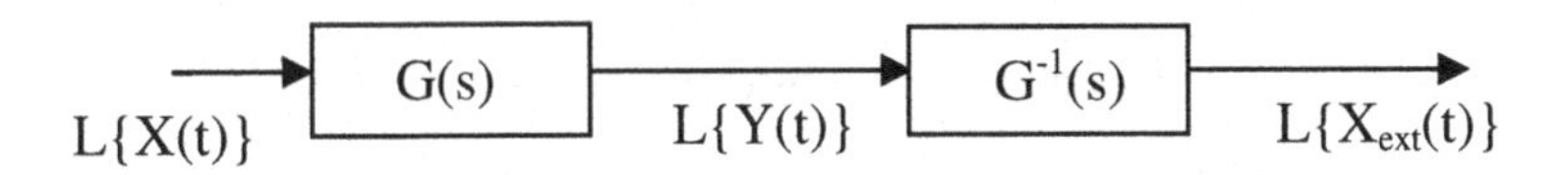

Fig. 4.21 Transfer functions for dynamic calibration

The effect of measurement and system noise are ignored in the first part of the analysis, but will be included in Ch. 4.4.

4.4.1 *Calibration for Computer-Based First Order Instruments*

For a simple introduction to the issues of dynamic calibration of sensors, consider a first order instrument

$$G(s) = \frac{k}{(1+T \cdot s)}$$

where k = gain, T = time constant.

The diagram of the system for static calibration is shown in Fig. 4.22, where an anti-aliasing filter is included to avoid sampling problems as a result of Analog to Digital Conversion (ADC)

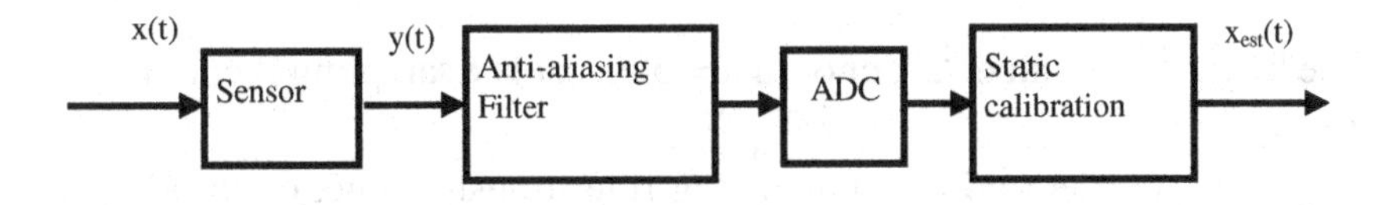

Fig. 4.22 Block diagram of the system with anti-aliasing filter

Figure 4.23 shows the corresponding transfer functions

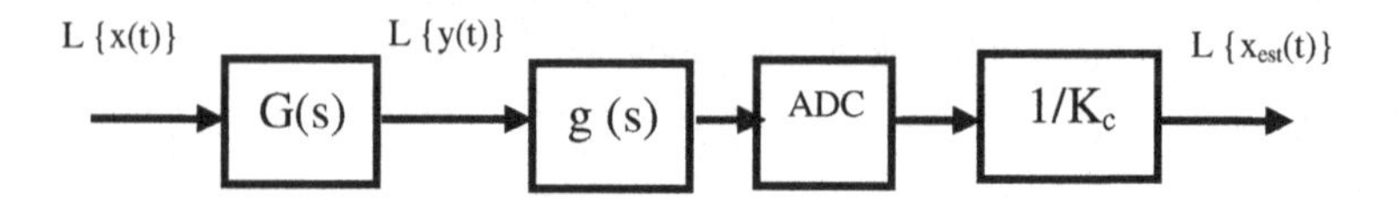

Fig. 4.23 Block diagram of transfer functions for Fig. 4.21

where, for static calibration, $K_c = G^{-1}(j \cdot 0)$.

For numerical illustration, the parameters of G(s) are chosen $k = 5 = K_c$ and $T = 0.01$ such that the bandwidth frequency is $\omega_b \approx 100$. The sampling rate is chosen $\omega_s = 2000$ and, in order to satisfy the condition $\omega_s > 20 \cdot \omega_b$ [10], a first order anti-aliasing analog filter is required:

$$g(s) = \frac{a}{(s+a)}$$

such that:

$$g(j\omega) = \frac{a}{(a+j\omega)} = \frac{a(a-j\omega)}{(a^2+\omega^2)}$$

$$| g(s) |= \frac{a}{(a^2+\omega^2)^{1/2}}$$

$$\varphi = \tan^{-1}(-\omega/a)$$

The break point value “a”, chosen a = 500, satisfy sampling theorem $a < \omega_s/2$.

The first order anti-aliasing analog filter transfer function g(s) has no practical effect on the amplitude up to $\omega_b \approx 100$, while the phase lag at ω=100 is:

$$\varphi = \tan^{-1}(-\omega / a) = \tan^{-1}(-100 / 500) = -11.3^0$$

Consider a first order instrument

$$G(s) = k / (1+Ts)$$

where k = gain, T = time constant and

$$K_c = G^{-1}(j \cdot 0) = k$$

Example 4.10 MATLAB simulations use $k = 5 = K_c$, $T = 0.01$, *i.e.*

$$G(s) = 5 / (1 + 0.01s),$$
$$\omega_b \approx 100$$

Sampling rate ω_s has to be chosen at least 20 times higher than the bandwidth ω_b [70]

$$\omega_s > 20\ \omega_b.$$

A first order anti-aliasing analog filter

$$g(s) = a / (s + a)$$
$$g(j \cdot \omega) = a / (a + j \cdot \omega) = a \cdot (a - j\omega) / (a^2 + \omega^2)$$
$$|g(s)| = a / (a^2 + \omega^2)^{1/2}$$
$$\varphi = \tan^{-1} (- \omega / a)$$

where the break point value a has to be smaller than $\omega_s / 2$ to satisfy sampling theorem.

$$a < \omega_s / 2$$
$$\omega_s = 25\ \omega_b = 2500$$
$$a = \omega_s / 5 = 500 > \omega_b \approx 100$$

Bode diagram for anti-aliasing filter are obtained with the MATLAB program

```
a=500;
num=[0 a];
den=[1 a];
bode(num,den);grid
```

The results are shown in Fig. 4.24. This shows that the first order anti-aliasing analog filter has no practical effect on the amplitude up to $\omega_b \approx 100$, but the phase lag is already significant and decreases further with ω down to - 90^0.

4.4.2 *Phase Lead Compensation*

The phase lag due to the anti-aliasing analog filter can become a problem in multi-sensor measurements, due to different phase lags for different sensors [61, 62, 63, 66, 67]. To solve this problem, additional phase lead digital compensation is required. For example, in this case, the phase lead compensator is:

$$C(s) = \frac{(b/a)(s+a)}{(s+b)}$$

where $b > a$, [70]. This phase lead digital compensator can be written as:

$$C(s) = \frac{(s+a)}{a}\frac{b}{(s+b)}$$

which shows that it consists of an inverse part of the anti-aliasing analog filter:

$$g^{-1}(s) = \frac{(s+a)}{a}$$

and a low pass filter $b / (s + b)$ with cross over frequency $b > a$.

The phase lead digital compensator is based on

$$C(s) = \frac{(b/a)(s+a)}{(s+b)}$$

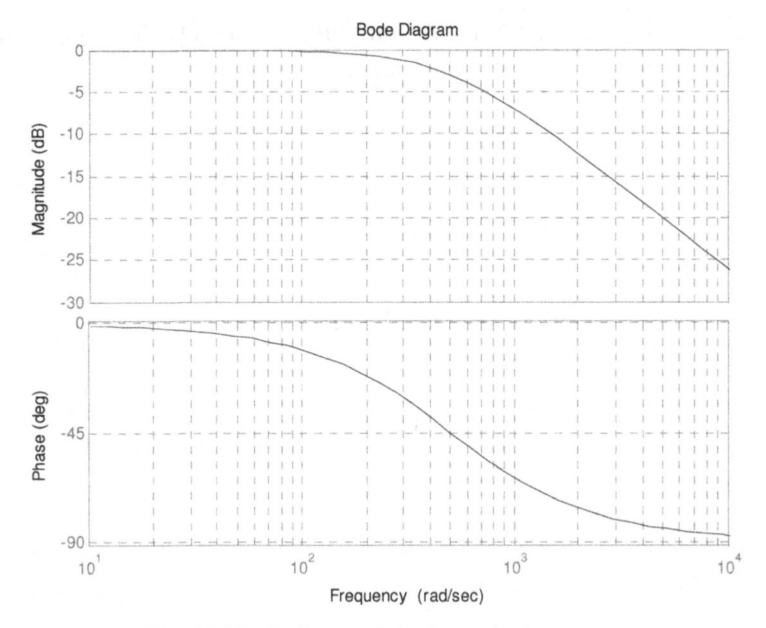

Fig. 4.24 Bode diagram for a first order instrument

Example 4.11 MATLAB program calculates the Bode diagram of the phase lead digital compensator with b = 1000 and a = 500

$$C(s) = \frac{2s+1000}{s+1000}$$

```
num=[2 1000];
den=[1 1000];
bode(num,den); grid
```

The result is shown in Fig. 4.25.

These results indicate that the phase lead compensator produces an increasing phase lead up to $\omega = 800$.

The anti-aliasing analog filter g(s) combined in series with the phase lead digital compensation C(s) has the transfer function:

$$g(s)\cdot C(s)=\frac{a}{(s+a)}\left(\frac{b}{a}\right)\frac{(s+a)}{(s+b)}=\frac{b}{(s+b)}$$

$$|g(s)C(s)|=\frac{b}{(b^2+\omega^2)^{1/2}}$$

$$\varphi=\tan^{-1}(-\omega/b)$$

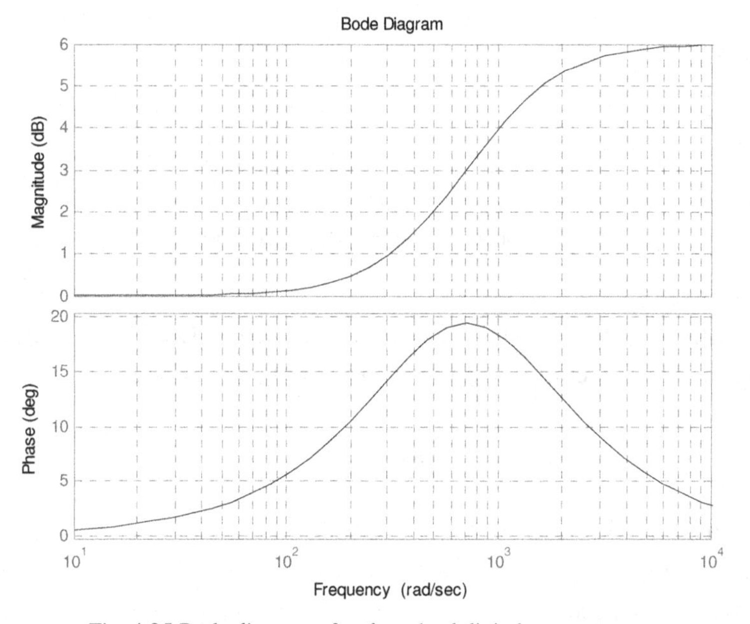

Fig. 4.25 Bode diagram of a phase lead digital compensator

Anti-aliasing analog filter g(s), combined in series with the phase lead digital compensation C(s), has the following transfer function

$$g(s)\cdot C(s)=\frac{a}{(s+a)}\left(\frac{b}{a}\right)\frac{(s+a)}{(s+b)}=\frac{b}{(s+b)}$$

Example 4.12 Bode diagram for the anti-aliasing is obtained with a MATLAB program for Bode diagram of g(s) · C(s) for b = 1000 > a = 500.

MATLAB program is

```
b=1000;
num=[0 b];
den=[1 b];
bode(num,den); grid
```

The results are shown in Fig. 4.26.

Bode diagram for the g(s) · C(s) shows that, compared to the results for g(s) in Fig. 4.24, in Fig. 4.26 the magnitude is maintained at 0 [dB] up to $\omega \approx 500$, and the phase lag is

$$\varphi = \varphi = \tan^{-1}(-\,\omega / b) \;= \varphi = \tan^{-1}(-\,\omega /1000)$$

the phase decreases slowly with ω.

At ω = 100 the phase is - 5.7 [^{0}], which, as a result of the phase lead compensation C(s), is a reduction for g(s) · C(s) to half of the phase lag of g(s).

The overall transfer function, including the transfer functions of the first order instrument and the anti-aliasing filter, ignoring the effect of ADC, is

$$L\{x_{est}(t)\} / L\{x(t)\} = G(s) \cdot g(s) \cdot C(s) / K_c = [k / (1 + Ts)] \cdot [a / (s + a)]$$
$$(b / a) \, [(s + a) / (s + b)] / k$$

In this case of assumed exact cancellation of k · a /(s + a)

$$L\{\ x_{est}(t)\ \} / L\{x(t)\} = b / [(1 + Ts) \cdot (s + b)] = 1/ [(1 + Ts) \cdot (1 + s / b)]$$

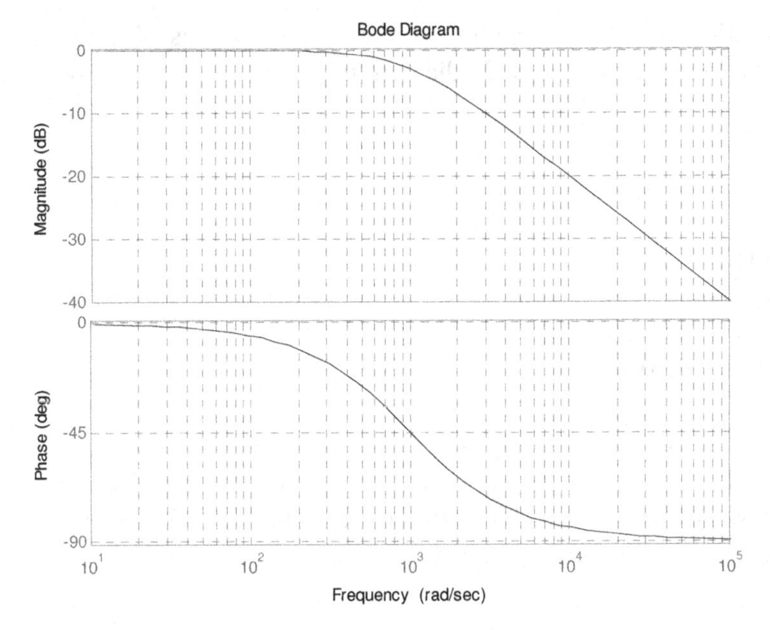

Fig. 4.26 Bode diagram for anti-aliasing analog filter and phase lead digital compensation

For T = 0.01 and b = 1000

$$L\{x_{est}(t)\} \,/\, L\{x(t)\} = 1 \,/\, (1+0.01 \cdot s) \cdot (1 + s \,/\, 1000)$$

or

$$L\{x_{est}(t)\} \,/\, L\{x(t)\} = 1 \,/\, (0.00001 \cdot s^2 + 0.011 \cdot s + 1)$$

Example 4.13 MATLAB program for the Bode diagram is

```
num=[0 0 1];
den=[0.00001 0.011 1];
bode(num,den); grid
```

The results are shown in Fig. 4.27.

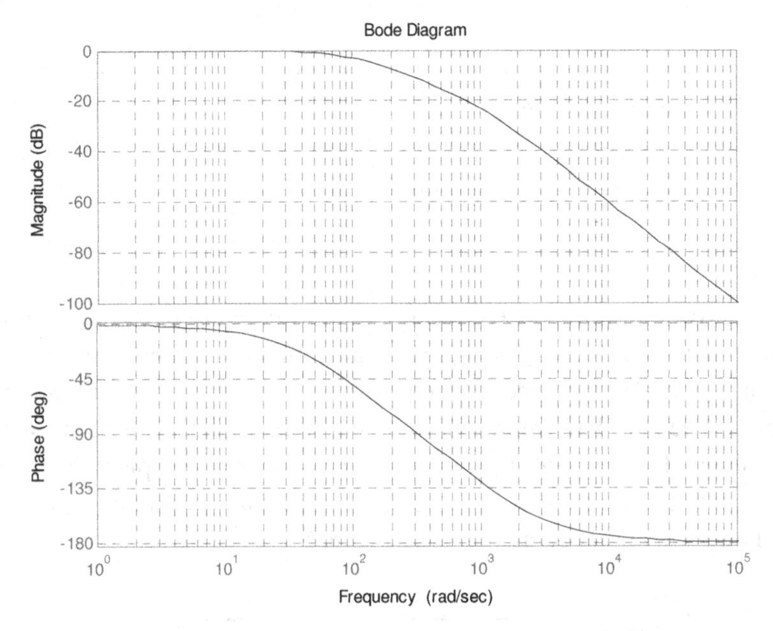

Fig. 4.27 Bode diagram for the first order instrument with the anti-aliasing filter and phase lead compensation

The results show that the effect of anti-aliasing filter combined with the phase lead digital compensator is a slight increase of bandwidth and a reduction of the phase shift. The anti-aliasing analog filter is needed to remove frequencies that can generate alias frequencies after ADC.

In multi-sensor measurements, due to different phase lags for different sensors, phase lead compensation might be required [8, 9]. For instance, two first order sensors with transfer functions

$$\frac{1}{1+T_1 s}$$

and

$$\frac{1}{1+T_2 s}$$

where the time constants T_1 and T_2 have different values, such that the phase lags are also different,

$$\phi_1 = \tan^{-1}(-T_1\omega)$$
$$\phi_2 = \tan^{-1}(-T_2\omega)$$

i.e. a relative phase lag difference Φ_1 - Φ_2 with regard to the same input. As shown above, the proposed phase lead compensation contains already dynamic compensation and, consequently, a separate static calibration is no more needed. This represents an acceptable solution for first order instruments. Higher order instruments require a more complex phase lead compensation of the anti-aliasing filter with the dynamic calibration based on inverse model, as shown in Fig. 4.28.

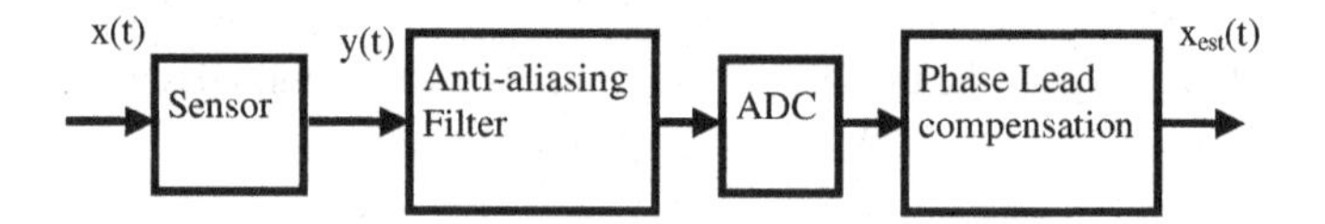

Fig. 4.28 Block diagram of the phase lead compensation and the anti-aliasing filter

A low pass filter to limit computations of inverse dynamics below very high frequencies range that can lead to overflow in numerical computations, low signal to noise ratio etc.

4.4.3 *Full and Reduced Order Dynamic Compensators*

Consider the transfer function G(s) of the high order instruments with P poles and $Z \leq P$ zeros:

$$G(s) = \frac{(s^z + b_{z-1}s^{z-1} + ... + b_0)}{(s^p + a_{p-1}s^{p-1} + ... + a_0)}$$

For G(s) with Z < P, $G^{-1}(s)$ with p = Z poles and z = P poles, and Z < P for G(s) results in z > p for $G^{-1}(s)$. Consequently, $G^{-1}(s)$ can be separated into a polynomial of order z - p in s and a rational function with the same number of poles and zeros equal to p. The polynomial of order z - p will operate on the input to $G^{-1}(s)$ as a high z - p order derivative. In the ideal case of measurements without noise, $G(s) \cdot G^{-1}(s) = 1$ and ideal dynamic calibration is obtained. In reality, noisy measurement occur and signal input to dynamic calibration using $G^{-1}(s)$, the high order derivative of random high frequency measurement noise has a very severe effect on calibration, in fact resulting in an ill-posed inverse problem. Dynamic calibration $G(s) \cdot G^{-1}(s)$ is not desirable in this case and reduced order inverse dynamics $G_r^{-1}(s)$ intends to remedy this effect.

Assume that only some of the poles and zeros have imaginary part within a frequency domain of interest ω_{use} and that the phase lead digital compensation is set up such that the poles of zeros outside this domain can be ignored in dynamic compensation. In this case, a reduced order $G_r^{-1}(s)$ can be used. This approach was first implemented with operational amplifiers that limited the domain of applications [71]. At this time, dynamic compensation can be implemented on embedded digital hardware which can be collocated with the instruments. For example, if only two complex conjugate poles, $-a_1 \pm j \cdot \omega_1$, $-a_2 \pm j \cdot \omega_2$ and two complex conjugate zeros, $-a_3 \pm j \cdot \omega_3$, $-a_4 \pm j \cdot \omega_4$, fall in this domain 0 - ω_{use}, steady state unit step response of $G(s) \cdot G_r^{-1}(s)(1 / s)$ has actually a constant value for

$$\lim_{s\to 0} s \cdot G(s) \cdot G_r^{-1}(s) \cdot \frac{1}{s} = \lim_{s\to 0} \frac{(s^{z-4} + b_{z-5}s^{z-5} + ... + b_0)}{(s^{p-4} + a_{p-4}s^{p-5} + ... + a_0)} = \frac{b_0}{a_0}$$

In this case, unit step response $G(s) \cdot G_r^{-1}(s)$ tends towards a_0/b_0 for frequencies $\omega < \omega_{use}$.

Similar to the previous section, an analog anti-aliasing filter and digital phase lead compensation are still needed for avoiding alias frequencies after ADC and for correcting the phase leg introduced by the anti-aliasing filter.

Example 4.14 In case that only two complex conjugate zeros and two complex conjugate poles for in the domain 0 to ω_{use}, phase lag digital compensation requires only a second order polynomial as numerator and another second order polynomial as denominator [71].

A generic sensor transfer function can be written as follows:

$$G(s)=\frac{(s+a_3+j\omega_3)(s+a_3-j\omega_3)(s+a_4+j\omega_4)(s+a_4-j\omega_4)(s^{z-4}+b_{z-5}s^{z-5}+..+b_0)}{(s+a_1+j\omega_1)(s+a_1-j\omega_1)(s+a_2+j\omega_2)(s+a_2-j\omega_2)(s^{p-4}+a_{p-4}s^{p-5}+..+a_0)}$$

where it is assumed that the two pairs of complex conjugate zeros the two pairs of complex conjugate poles have:

$$\omega_i < \omega_{use} \text{ for } i = 1,2,3,4$$

while all other complex zeros and poles have frequencies larger than ω_{use}

$$\omega_i > \omega_{use} \text{ for } i > 5$$

In this case, the dynamic compensator has to cancel only $\omega_i < \omega_{use}$, *i.e.* a reduced order compensator with the transfer function can be used:

$$G_r^{-1}(s)=\frac{(s+a_1+j\omega_1)(s+a_1-j\omega_1)(s+a_2+j\omega_2)(s+a_2-j\omega_2)}{(s+a_3+j\omega_3)(s+a_3-j\omega_3)(s+a_4+j\omega_4)(s+a_4-j\omega_4)}$$

such that

$$G(s)\cdot G_r^{-1}(s)=\frac{(s^{z-4}+b_{z-5}s^{z-5}+...+b_0)}{(s^{p-4}+a_{p-4}s^{p-5}+...+a_0)}$$

where $G(s)\cdot G_r^{-1}(s)$ has complex zeros and poles with frequencies $\omega > \omega_{use}$.

4.4.3.1 *First order instrument*

Inverse transfer function for a first order instrument is

$$G^{-1}(s) = (1+Ts)/ k$$

The block diagram for the dynamically calibrated first order sensor is shown in Fig. 4.29.

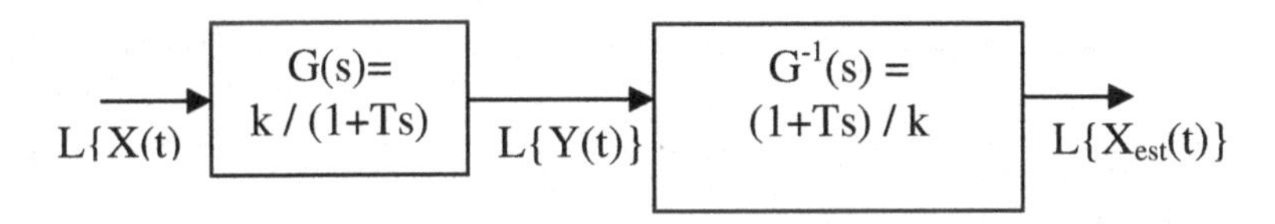

Fig. 4.29 The block diagram for the dynamically calibrated first order sensor

Bode diagram permits to identify the effect of $G^{-1}(s)$ on the estimation of the input.

Example 4.15 Bode diagram of the first order instruments compensator is obtained for k = 5 and T = 0.01.

MATLAB program is

```
num=[T 1];
den=[0 k];
bode(num,den);grid;
```

The results are shown in Fig. 4.30.

The magnitude of the inverse dynamics compensator $|G^{-1}(s)| = |(1 + Ts) / k|$ shows 20 dB/decade increase beyond bandwidth cutoff frequency, $\omega_b = 100$, indicating that growing computational difficulties can occur $\omega >> \omega_b$.

4.4.3.2 *Second order instrument*

Assume second order instrument transfer function

$$G(s) = k / (s^2 + b \cdot s + c)$$

Dynamic calibration in this case is achieved by

$$G^{-1}(s) = (s^2 + b \cdot s + c) / k$$

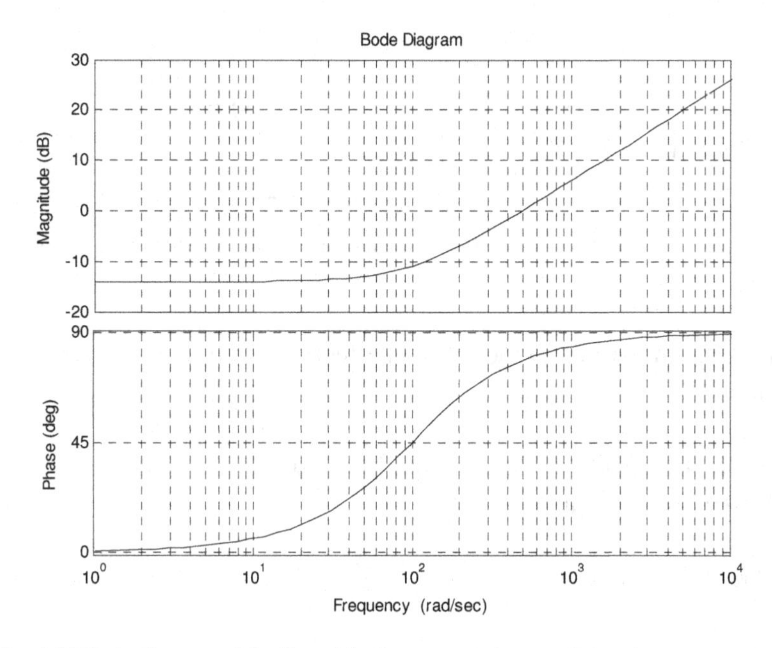

Fig. 4.30 Bode diagram of the first order instruments inverse dynamics compensator

Example 4.16 MATLAB program for $G^{-1}(s)$ is

```
k=1;
b=12;
c=100;
den =[0 0 k];
num =[1 b c];
bode(num,den);
grid;
```

The results are shown in Fig. 4.31.

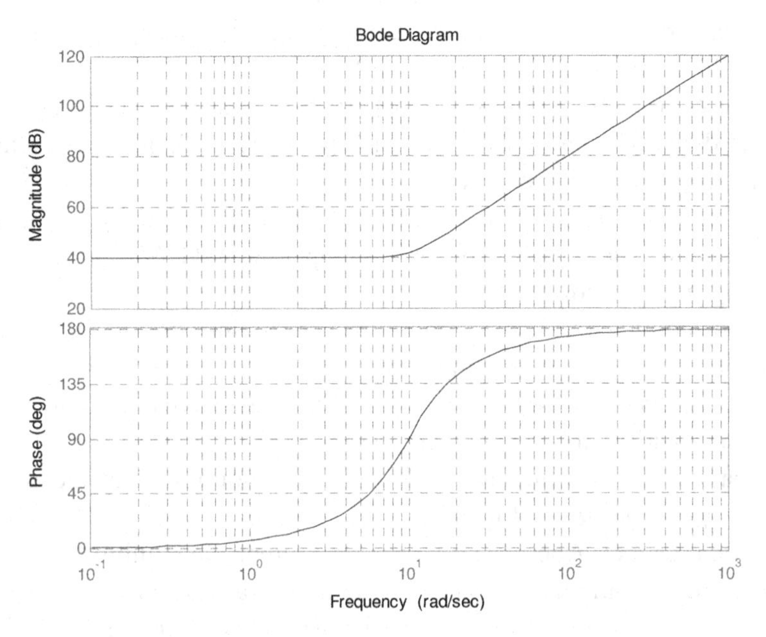

Fig. 4.31 Bode diagram of $G^{-1}(s) = s^2+12 \cdot s + 100$

The magnitude of the inverse dynamic compensator $|G^{-1}(s)|$ shows 40 dB/decade increase beyond bandwidth cutoff frequency, $\omega_b = 10$, indicating growing computational difficulties as $\omega >> \omega_b$, even more significant in the case of first order instruments. N-order instruments will have 20 N dB/decade increase beyond bandwidth cutoff frequency, *i.e.* more significant increase of Magnitude values beyond bandwidth cutoff frequency.

Some solutions to the above difficulties were outlined for the case of first order instruments:

-low pass filter for removing high frequency noise in the sensor output
-reduced order inverse dynamics

-Modified Output Approach (MOA) applied to non-minimum phase systems to avoid unstable inverse dynamics.

In general, inverse dynamic compensator for increasing sensors bandwidth requires to solve various difficulties:

-computational difficulties as $\omega >> \omega_b$ due to increasing magnitude of the inverse dynamic compensator $|G^{-1}(s)|$ = for $\omega > \omega_b$; digital word length limitation can lead to overflow;

-high frequency noise in the sensor output is amplified by increasing magnitude of the inverse dynamic compensator $|G^{-1}(s)|$ for $\omega > \omega_b$ reducing signal to noise ratio;

-un-modeled dynamics and parametric uncertainty result in reduced effect of inverse dynamics compensator;

-non-minimum phase systems have unstable inverse dynamics [74].

Some solutions to the above difficulties are:

-low pass filter for removing high frequency noise w(t) in the sensor output $y_m(t)$, as shown in Fig. 4.32

-Modified Output Approach (MOA) or Output Redefinition Method, applied to non-minimum phase systems to avoid unstable inverse dynamics [74], shown in Fig. 4.33. This method is applied in Ch. 9. The result is effective for frequencies lower than the positive zero and not to higher frequencies.

In practice the increase beyond bandwidth cutoff frequency, ω_b is normally limited up to a maximum useful frequency component ω_{use}. Limiting inverse dynamic compensator to $\omega < \omega_{use}$ avoids reaching unacceptable high magnitudes of the inverse dynamic compensator, making this approach of interest for computer based instrumentation.

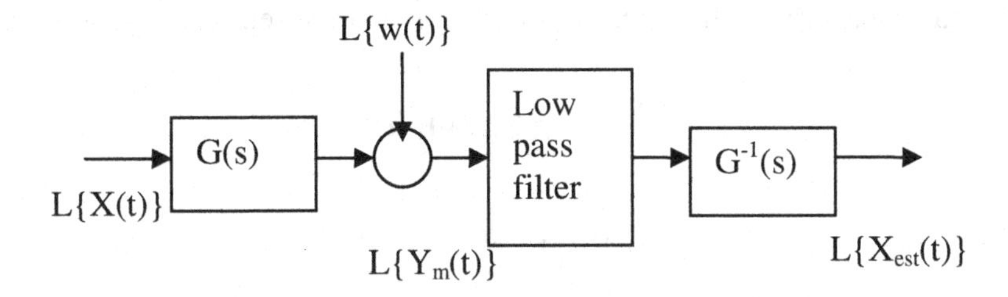

Fig. 4.32 Low pass filter for removing high frequency noise

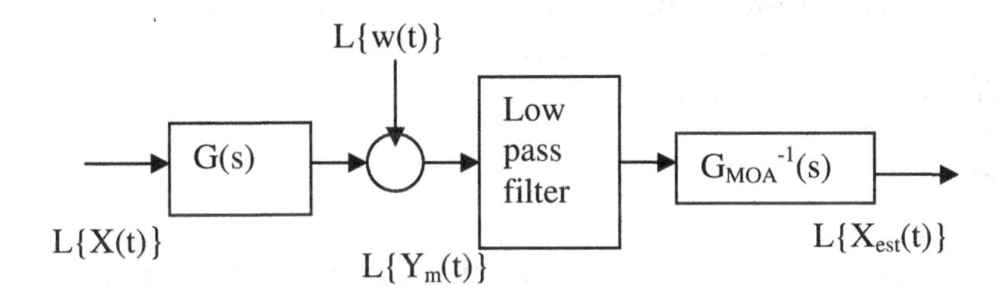

Fig. 4.33 Modified Output Approach

4.5 Dynamic Calibration in Case of Noisy Measurements

Inverse problems for Linear Time Invariant (LTI) systems can be formulated for different representations of the forward model of the system, different for lumped parameters systems from distributed parameters systems.

Inverse problems for LTI lumped parameters models of sensors can be formulated for different representations of the forward model of the system:

A) state space representation,

B) complex functions representation,

C) convolution integral representation,

D) matrix form representation.

A) State space representation in case of noisy measurements is given by:

$$\frac{d\mathbf{X}(t)}{dt} = \mathbf{a} \cdot \mathbf{X}(t) + \mathbf{b} \cdot \mathbf{U}(t)$$

$$\mathbf{Y}_m(t) = \mathbf{c} \cdot \mathbf{X}(t) - \mathbf{W}(t)$$

where:

X [N ·1] is the system state vector
U [M ·1] is the system input vcctor
Y [P ·1] is the system output vector,
W(t) is measurement noise vector
M≤N
P≤N

This model is used to obtain the estimation $\mathbf{U}_{est}$ of **U** given $\mathbf{Y}_m$ and sensor model.

Denote noiseless output

$$\mathbf{Y}(t) = \mathbf{Y}_m(t) - \mathbf{W}(t)$$

such that the output equation becomes

$$\mathbf{Y}(t) = \mathbf{c} \cdot \mathbf{X}(t)$$

The inverse problem of estimating the input $\mathbf{U}_{est}(t)$ from measurements y(t) is obtained solving output equation using the generalized inverse $\mathbf{c}^{-g}$

$$\mathbf{X}(t) = \mathbf{c}^{-g}\mathbf{Y}(t)$$

$$\frac{d\mathbf{X}(t)}{dt} = \mathbf{c}^{-g}\frac{d\mathbf{y}(t)}{dt}$$

that gives:

$$\mathbf{c}^{-g}\,\frac{d\mathbf{y}(t)}{dt} = \mathbf{a}\cdot\mathbf{c}^{-g}\mathbf{Y}(t) + \mathbf{b}\cdot\mathbf{U}(t)$$

The solution for $\mathbf{U}(t)$ of the state equation is

$$\mathbf{U}_{est}(t) = \mathbf{b}^{-g}\left(\frac{d\mathbf{X}(t)}{dt} - \mathbf{a}\cdot\mathbf{X}(t)\right)$$

or:

$$\mathbf{U}_{est}(t) = \mathbf{b}^{-g}\left(\frac{c^{-g}d\mathbf{Y}(t)}{dt} - \mathbf{a}\cdot\mathbf{c}^{-g}\cdot\mathbf{Y}(t)\right)$$

or, taking into account the noisy output

$$\mathbf{Y}_m(t) = \mathbf{Y}(t) + \mathbf{W}(t)$$

$$\mathbf{U}_{est}(t) = \mathbf{b}^{-g}\left(\frac{\mathbf{c}^{-g}d([\mathbf{Y}(t)-\mathbf{W}(t)]}{dt} - \mathbf{a}\cdot\mathbf{c}^{-g}[\mathbf{Y}(t)-\mathbf{W}(t)]\right)$$

This solution requires the calculation of generalized inverses $\mathbf{b}^{-g}$ and $\mathbf{c}^{-g}$ as well as the derivative $d\mathbf{Y}(t)/dt$. Real-time implementation of this solution might be computationally intensive and requires specific code for each application. The presence of fast varying noise $\mathbf{W}(t)$ might lead to very low signal to noise ratios that reduces in this case the usefulness of dynamic calibration.

B) Complex functions representation is obtained from the Laplace transform of the state space equations for zero initial conditions

$$(\mathbf{I}\cdot s - \mathbf{a})\cdot\mathbf{X}(s) = \mathbf{b}\cdot\mathbf{U}(s)$$
$$\mathbf{Y}(s) = \mathbf{c}\cdot\mathbf{X}(s)$$

For $\mathbf{X}(s) = \mathbf{c}^{-g}\cdot\mathbf{Y}(s)$, state equation, after eliminating $\mathbf{X}(s)$, becomes

$$(\mathbf{I}\cdot s - \mathbf{a})\cdot\mathbf{c}^{-g}\cdot\mathbf{Y}(s) = \mathbf{b}\cdot\mathbf{U}(s)$$

Solving this equation algebraically for $\mathbf{U}(s)$ gives

$$\mathbf{b}^{-g}(\mathbf{I}\cdot s - \mathbf{a})c^{-g}\mathbf{Y}(s) = \mathbf{U}(s)$$

The estimate $\mathbf{U}_{est}(s)$ results as follows:

$$\mathbf{U}_{est}(s) = \mathbf{b}^{-g} \cdot (\mathbf{I}s - \mathbf{a}) \cdot \mathbf{c}^{-g} \cdot \mathbf{Y}(s)$$

or

$$\mathbf{U}_{est}(s) = \mathbf{b}^{-g} \cdot (\mathbf{I} \cdot s - \mathbf{a}) \cdot c^{-g} \cdot [\mathbf{Y}(s) - \mathbf{W}(s)]$$

As expected, this solution requires also the calculation of generalized inverses $\mathbf{b}^{-g}$ and $\mathbf{c}^{-g}$. Moreover, in the feed-forward path of the sensor-dynamic compensator, the presence of "s" indicates the same requirement for the time derivative. Real-time implementation of this complex function solution is not desirable.

C) Convolution integral representation is of interest as a link to non-linear forward problems formulation using integral equations and as a basis for developing computationally efficient matrix formulation.

The principle of superposition, valid for linear systems, gives [70]

$$Y(t) = \int_{-\infty}^{\infty} U(\tau) \cdot h(t, \tau) d\tau$$

where $h(t, \tau)$ is the impulse response of the system, for the impulse assumed applied at any time τ. In the case of LTI systems,

$$h(t, \tau) = h(t - \tau)$$

i.e. it depends only on the difference between the time τ when the impulse is applied and the time t when the response y is observed. This property greatly reduces the computation of the impulse response h. The convolution integral for LTI systems is:

$$Y(t) = \int_{-\infty}^{\infty} U(\tau) \cdot h(t - \tau) d\tau$$

The calculation of the impulse response h for LTI system results from considering a unit impulse input U(t) = δ(t), such that:

$$\frac{d\mathbf{x}(t)}{dt} = \mathbf{a} \cdot \mathbf{X}(t) + \mathbf{b} \cdot \boldsymbol{\delta}(t)$$

and:

$$\mathbf{c}^{-g} \frac{d\mathbf{Y}(t)}{dt} = \mathbf{a} \cdot \mathbf{c}^{-g} \cdot \mathbf{Y}(t) + \mathbf{b} \cdot \boldsymbol{\delta}(t)$$

Rather than solving analytically this equation, complex functions representation can be used to obtain the transfer function. In scalar case

$$h(s) = Y(s) / \delta(s)$$

For the unit impulse input U(s) = δ(s) = 1, impulse response h(s) can be calculated in time domain, h(t), using the inverse Laplace transform L^{-1} that gives $h(t) = L^{-1} \{h(s)\}$. Convolution integral can be reformulated in the discrete form of a convolution sum using shifted impulse response h_{i-j} for the sampled time interval t - τ with sampling period T_s, such that the discrete time τ_j, (corresponding to the continuous time τ, when the impulse is applied [67]), where $\tau_j = \tau / T_s$ and the time t_i when the response y is observed $t_j = t / T_s$, such that t - τ in discrete time is $(i - j)T_s$ or i - j in steps.

Convolution sum for LTI discrete systems

$$Y_i = \sum_{k=-\infty}^{\infty} u_k \cdot h_{i-k}$$

corresponds to:

$$y(t) = \int_{-\infty}^{\infty} U(\tau) \cdot h(t-\tau)\, d\tau$$

For input signals of limited duration and/or damped systems this can be written, after discretization, in matrix form, as shown in the next section.

D) Matrix form representation of the forward model, in discrete time, is:

$$\mathbf{y} = \mathbf{h} \cdot \mathbf{u}$$

where **u** is input $[N_u \cdot 1]$ vector, **y** is output $[N_y \cdot 1]$ vector and **h** is $[N_u \cdot N_y]$ matrix.

It has to be taken into account that **y** and **u** in matrix form representation and **Y** and **U** in state space representation have different contents.

Inverse model permits to calculate the estimate of the sensor input signal:

$$\mathbf{u}_{est} = \mathbf{h}^{-1} \cdot \mathbf{y}$$

In fact noiseless output y is not available. Replacing $\mathbf{y} = \mathbf{y}_m - \mathbf{W}$

$$\mathbf{u}_{est} = \mathbf{h}^{-1} \cdot (\mathbf{y} - \mathbf{W})$$

This estimation requires $\mathbf{y}_m$, which is a noisy signal, as well as the knowledge of the random noise characteristics. Ignoring the noise, the above equation gives an approximate estimation $\mathbf{u}_{est}$.

The inversion of **h**, a $[N_u \cdot N_y]$ matrix, can only be obtained as a pseudo-inverse $\mathbf{h}^{-g}$. This is a typical difficulty in inverse problem solving, that is extensively investigated in the specialized literature [67-73].

As a result, the proposed dynamic calibration approach for sensors can benefit from the existing methods for obtaining numerical solutions for inverse problems.

In fact, SVD and regularization methods presented in Ch. 3, as well as reduced order dynamics method, presented in Ch. 4.4.3, limit the solutions to inverse problems to lower frequencies domains to avoid the effect of high frequency measurement noise.

4.6 State Estimation for Indirect Sensing

4.6.1 *Derivation of the Estimator for Indirect States Estimation Using Matrix Inversion Approach*

So far in Ch. 4 sensor calibration problem was formulated for sensor input estimation $\mathbf{U}_{est}(t)$ from given sensor output measurements $\mathbf{y}(t)$. In this section, the problem is the estimation of states that are not directly measured, *i.e.* the case of states that do not provide direct inputs to sensors. The approach is based on the dynamic model linking measured states to the other states that are not directly measured.

Assume a LTI (Linear Time Invariant) system in state-space representation

$$d\mathbf{X}(t) / dt = \mathbf{a} \cdot \mathbf{X}(t) + \mathbf{b} \cdot \mathbf{U}(t)$$
$$\mathbf{Y}(t) = \mathbf{c} \cdot \mathbf{X}(t)$$

where

$\mathbf{X}$ $[N_x \cdot 1]$ is the system state vector
$\mathbf{U}$ $[N_u \cdot 1]$ is the system input vector
$\mathbf{Y}$ $[N_y \cdot 1]$ is the system output vector,

where $N_u \leq N_x$ and $N_y \leq N_x$.

Output matrix $\mathbf{c}$ is assumed to distinguish between n_1 measured states $\mathbf{X}_1$ $[N_{x1} \cdot 1]$ (directly measured by n_1 sensors), from the n-n_1 states, intended to monitor, $\mathbf{X}_2$ $[(N_x - N_{x1}) \cdot 1]$ not measured directly by sensors [72]. Indirect measurement of systems defined by ODE models are a simplified case of non-collocated measurements of systems defined by PDE and, for this reason, in both cases they will be named here as cases of non-collocated sensing, even if in systems defined by ODE, space variable is not present. In this context, indirectly measured states are linked by lumped parameters models to directly measured states. In this case, the state vector is partitioned as follows

$$\mathbf{X} = \left[\frac{\mathbf{X}_1}{\mathbf{X}_2}\right]$$

System state space model can be partitioned as well

$$\begin{aligned} d\mathbf{X}_1(t)/dt &= \mathbf{a}_{11}\cdot\mathbf{X}_1(t) + \mathbf{a}_{12}\cdot\mathbf{X}_2(t) + \mathbf{b}_1\cdot\mathbf{U}(t) \\ d\mathbf{X}_2(t)/dt &= \mathbf{a}_{21}\cdot\mathbf{X}_1(t) + \mathbf{a}_{22}\cdot\mathbf{X}_2(t) + \mathbf{b}_2\cdot\mathbf{U}(t) \\ \mathbf{Y}(t) &= \mathbf{c}_1\cdot\mathbf{X}_1(t) \end{aligned}$$

and $\mathbf{c}_2$ has only zero elements such that $\mathbf{c}_2 \cdot \mathbf{X}_2=0$.

Indirect sensing requires to calculate estimates $\mathbf{X}_{2est}$ given direct measurements $\mathbf{Y}(t)$ of only $\mathbf{X}_1(t)$.

Sensors input vector $\mathbf{u}$ $[n_u \cdot 1]$ is system output vector the states $\mathbf{X}_1$ directly measured by n_1 sensors

$$\mathbf{u}(t) = \mathbf{Y}(t) = \mathbf{c}_1\cdot\mathbf{X}_1(t)$$

Sensors output vector $\mathbf{y}(t)$ contains the only signals available to estimates $\mathbf{X}_{2est}$ for the indirectly measured states $\mathbf{X}_2$.

Assume a state-space representation of the sensors

$$\begin{aligned} d\mathbf{x}(t)/dt &= \mathbf{A}\cdot\mathbf{x}(t) + \mathbf{B}\cdot\mathbf{u}(t) \\ \mathbf{y}(t) &= \mathbf{C}\cdot\mathbf{x}(t) \end{aligned}$$

where

$\mathbf{x}$ $[n_x \cdot 1]$ is the sensor state vector

$\mathbf{u}$ $[n_u \cdot 1]$ is the sensor input vector, $n_u = N_y$ for $\mathbf{Y} = \mathbf{u}$

$\mathbf{y}[n_p \cdot 1]$ is the sensor output vector,

where $n_u \leq n_x$ and $n_p \leq n_x$.

Sensor output $\mathbf{y}$ is used for estimating directly $\mathbf{X}_1$ and indirectly $\mathbf{X}_2$. The block diagram is shown in Fig. 4.34.

Dynamic calibration for the sensors in complex domain gives

$$\mathbf{u}_{est}(s) = \mathbf{B}^{-g}\cdot(\mathbf{I}\cdot s - \mathbf{A})\cdot\mathbf{C}^{-g}\cdot\mathbf{y}(s)$$

Non-collocated system states X_2 result from the system state equation

$$d\mathbf{X}_2(t)/dt = \mathbf{a}_{21}\cdot\mathbf{X}_1(t) + \mathbf{a}_{22}\cdot\mathbf{X}_2(t) + \mathbf{b}_2\cdot\mathbf{U}(t)$$

Estimation $\mathbf{X}_{2est}$ of the system states $\mathbf{X}_2$ cam be obtained with a reduced order observer [72], in fact the inverse model

$$d\mathbf{X}_{2est}(t) / dt = \mathbf{a}_{21} \cdot \mathbf{X}_{1est}(t) + \mathbf{a}_{22} \cdot \mathbf{X}_{2est}(t) + \mathbf{b}_2 \cdot \mathbf{U}(t)$$

where $\mathbf{X}_{1est}$ (t) can be obtained from a matrix inversion $\mathbf{c}^{-1}$ and sensor output $\mathbf{Y}(t)$

$$\mathbf{X}_{1est}(t) = \mathbf{c}^{-1} \cdot \mathbf{Y}(t)$$

The matrix $\mathbf{c}[P * N]$ is rarely square and nonsingular and a pseudo-inverses or generalized inverse $\mathbf{c}^{-g}$ has to be used instead

$$\mathbf{X}_{1est}(t) = \mathbf{c}^{-g} \cdot \mathbf{Y}_{est}(t) = \mathbf{c}^{-g} \cdot \mathbf{u}_{est}$$

Observer dynamics for $\mathbf{X}_{2est}(t)$ is given by

$$d\mathbf{X}_{2est}(t) / dt - \mathbf{a}_{22} \cdot \mathbf{X}_{2\,est}(t) = \mathbf{a}_{21} \cdot \mathbf{c}^{-g} \cdot \mathbf{u}_{est} + \mathbf{b}_2 \cdot \mathbf{U}(t)$$

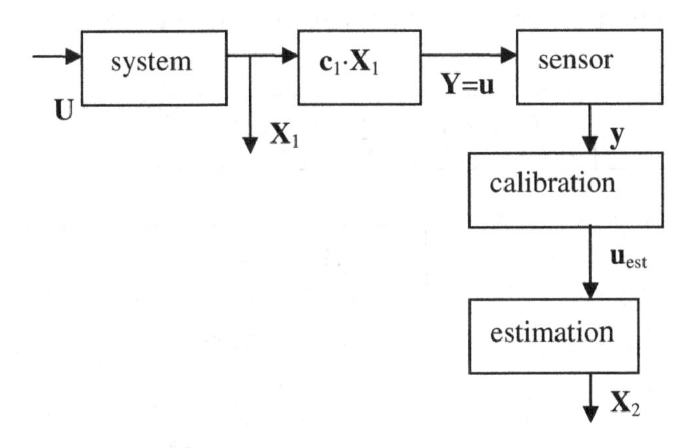

Fig. 4.34 Indirect state estimation

While $\mathbf{X}_{1est}(t)$ can be obtained from a matrix inversion $\mathbf{c}^{-1}$ and sensor output $\mathbf{Y}(t)$, the calculation of $\mathbf{X}_{2est}(t)$ requires the inputs $\mathbf{u}_{est}$ from sensor dynamic calibration and system input $\mathbf{U}(t)$.

After taking Laplace transform

$$(\mathbf{I}s - \mathbf{a}_{22}) \cdot \mathbf{X}_{2est}(s) = \mathbf{a}_{21} \cdot \mathbf{c}^{-g} \cdot \mathbf{u}_{est}(s) + \mathbf{b}_2 \cdot \mathbf{U}(s)$$

or, an observer based on a square matrix inversion

$$\mathbf{X}_{2est}(s) = (\mathbf{I} \cdot s - \mathbf{a}_{22})^{-1} \cdot [\mathbf{a}_{21} \cdot \mathbf{c}^{-g} \cdot \mathbf{u}_{est}(s) + \mathbf{b}_2 \cdot \mathbf{U}(s)]$$

or

$$\mathbf{X}_{2est}(s) = (\mathbf{I} \cdot s - \mathbf{a}_{22})^{-1} \cdot \mathbf{a}_{21} \cdot \mathbf{X}_{1est}(s) + (\mathbf{I} \cdot s - \mathbf{a}_{22})^{-1} \cdot \mathbf{b}_2 \cdot \mathbf{U}(s)$$

The block diagram is shown in Fig. 4.35. Digital computations of the estimates $\mathbf{X}_{1est}$ and $\mathbf{X}_{2est}$ consists of

$$\mathbf{X}_{1est}(s) = \mathbf{c}^{-g}(s) \cdot \mathbf{B}^{-g} \cdot (\mathbf{I} \cdot s - \mathbf{A}) \cdot \mathbf{C}^{-g} \cdot \mathbf{y}$$
$$\mathbf{X}_{2est}(s) = (\mathbf{I} \cdot s - \mathbf{a}_{22})^{-1} \cdot (\mathbf{a}_{21} \cdot \mathbf{X}_{1est}(s) + \mathbf{b}_2 \cdot \mathbf{U}(s))$$

This approach takes into account sensor dynamics and introduces dynamic calibration for the sensor, aspects often ignored in cases when sensors where assumed simply represented by the output matrix $\mathbf{c}$, i.e. for ideal sensors that measure collocated states exactly, $\mathbf{u}_{est} = \mathbf{u} = \mathbf{Y} = \mathbf{c}_1 \cdot \mathbf{X}_1$ [72].

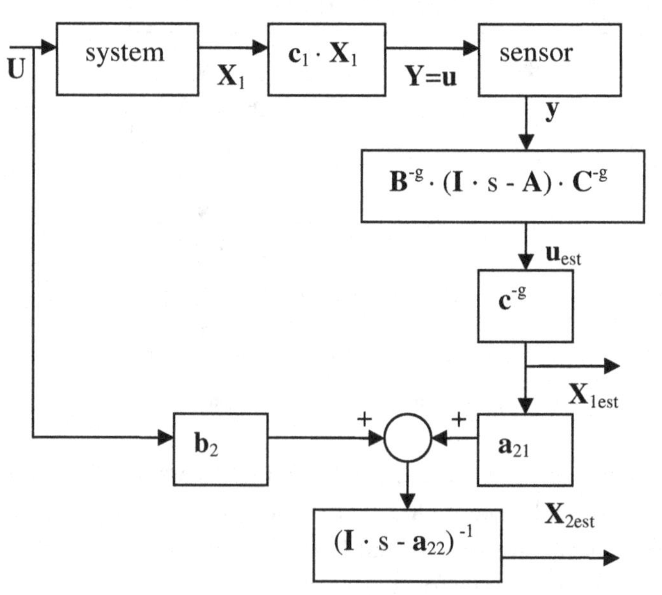

Fig. 4.35 Indirect state estimation in matrix form formulation

4.6.2 *Luenberger Observer and Kalman Filters*

It can be observed that besides the numerical difficulties mentioned for the dynamic calibration, the observer requires the calculation of eigenvalues of the square matrix ($\mathbf{I} \cdot s - \mathbf{a}_{22}$), that controls the rate of estimation convergence of $\mathbf{X}_{2est}$. [70].

A Luenberger observer might provide, however, a better control of the rate of estimation convergence for X_{2est} [70].

$$d\mathbf{X}_{1est}(t)/dt = \mathbf{a}_{11} \cdot \mathbf{X}_{1est}(t) + \mathbf{a}_{12} \cdot \mathbf{X}_{2est}(t) + \mathbf{b}_1 \cdot \mathbf{U}(t) + L_1 \cdot [\mathbf{Y}(t) - \mathbf{Y}_{est}(t)]$$
$$d\mathbf{X}_{2est}(t)/dt = \mathbf{a}_{21} \cdot \mathbf{X}_{1est}(t) + \mathbf{a}_{22} \cdot \mathbf{X}_{2est}(t) + \mathbf{b}_2 \cdot \mathbf{U}(t) + L_2 \cdot [\mathbf{Y}(t) - \mathbf{Y}_{est}(t)]$$
$$\mathbf{Y}(t) = \mathbf{c}_1 \cdot \mathbf{X}_1(t)$$
$$\mathbf{Y}_{est}(t) = \mathbf{c}_1 \cdot \mathbf{X}_{1est}(t)$$

where L_1 and L_2 are Luenberger observer gains that can be chosen such that the desired rate of estimation convergence is achieved.

In such a case, the overall calculation of $\mathbf{X}_{1est}$ and $\mathbf{X}_{2est}$ can be formulated as an Luenberger observer that would include not only system dynamics equations but also sensor dynamics equations. The advantage could be the avoidance of inclusion of signal derivatives in the feed-forward path. The disadvantage is the development of a centralized error [$\mathbf{Y}(t) - \mathbf{Y}_{est}(t)$] in a feedback approach on which relies the overall result, while the above developed approach includes decentralized sensor dynamic calibration adapted to the specific characteristics of the sensor. Moreover, Luenberger observer still requires the calculation of $\mathbf{c}^{-1}$, or if $\mathbf{c}$ is a singular matrix, of $\mathbf{c}^{-g}$. [72].

Kalman filters serve the same purpose, but include the effect of random noise. White noise assumption might be too restrictive, while colored noise assumption might result in a significant computation burden in real-time applications. In practical applications a choice might have to be made between a deterministic approach applied after filtering random noise and a stochastic approach in which random noise effects are included in the direct and inverse problem formulation [41, 45, 65, 67].

Indirect state estimation shown in Fig. 4.35 and Luenberger observer parallel the two types of methods presented in Ch. 3.3 for solving inverse problems, matrix inversion and iterative methods, respectively.

For a full state feedback system

$$\mathbf{U} = \mathbf{K} \cdot \mathbf{X} = [\mathbf{K}_1 \mid \mathbf{K}_2] \left[\frac{\mathbf{X}_1}{\mathbf{X}_2} \right] = \mathbf{K}_1 \cdot \mathbf{X}_1 + \mathbf{K}_2 \cdot \mathbf{X}_2$$

$$d\mathbf{X}_{2est}(t) / dt = a_{21} \cdot \mathbf{X}_{1est}(t) + \mathbf{a}_{22} \cdot \mathbf{X}_{2est}(t) + \mathbf{b}_2 \cdot \mathbf{U}(t) = \mathbf{a}_{21} \cdot \mathbf{X}_{1est}(t) +$$
$$\mathbf{a}_{22} \cdot \mathbf{X}_{2est}(t) + \mathbf{b}_2 \cdot (\mathbf{K}_1 \cdot \mathbf{X}_{1est} + \mathbf{K}_2 \cdot \mathbf{X}_{2est}) =$$
$$(\mathbf{a}_{21} + \mathbf{b}_2 \cdot \mathbf{K}_1) \cdot \mathbf{X}_{1est}(t) + (\mathbf{a}_{22} + \mathbf{b}_2 \cdot \mathbf{K}_2) \cdot \mathbf{X}_{2est}(t)$$

$$\mathbf{Y}_{est}(t) = \mathbf{u}_{est}(t) = \mathbf{c}_1 \cdot \mathbf{X}_{1est}(t)$$

In this case the estimation does not require any autonomous system input

$$\mathbf{U} = 0$$
$$d\mathbf{X}_{2est}(t) / dt = \mathbf{a}_{21} \cdot \mathbf{X}_{1est}(t) + \mathbf{a}_{22} \cdot \mathbf{X}_{2est}(t)$$

or

$$d\mathbf{X}_{2est}(t) / dt - \mathbf{a}_{22} \cdot \mathbf{X}_{2est}(t) = \mathbf{a}_{21} \cdot \mathbf{X}_{1est}(t)$$

Taking Laplace transform for zero initial conditions

$$(s \cdot \mathbf{I}_{22} - \mathbf{a}_{22}) \cdot \mathbf{X}_{2est}(s) = \mathbf{a}_{21} \cdot \mathbf{X}_{1est}(s)$$
$$\mathbf{X}_{2est}(s) = (s \cdot \mathbf{I}_{22} - \mathbf{a}_{22})^{-1} \cdot \mathbf{a}_{21} \cdot \mathbf{X}_{1est}(s)$$

or

$$\mathbf{X}_{2est}(s) = (s \cdot \mathbf{I}_{22} - \mathbf{a}_{22})^{-1} \cdot \mathbf{a}_{21} \cdot \mathbf{c}^{-g}(s) \cdot \mathbf{B}^{-g} \cdot (s \cdot \mathbf{I} - \mathbf{A}) \cdot \mathbf{C}^{-g} \cdot \mathbf{y}(s)$$

For this passive system ($\mathbf{U} = 0$), both estimations $\mathbf{X}_{1est}$ and $\mathbf{X}_{2est}$ are based only on the measurements $\mathbf{y}(s)$. For example, this would permit to calculate inner states of passive Mass-Spring-Damper networks from some particular nodes displacement measurements or inner states of passive electric R-L-C networks from some particular node voltage measurements.

4.6.3 *Indirect Estimation of States and Inputs for LTI ODE Systems Using Matrix Inversion*

The state space model from Ch. 4.5.1, partitioned in states directly $\mathbf{X}_1$ and indirectly $\mathbf{X}_{2\ \text{measured}}$, can also be used to estimates the unknown input $\mathbf{U}$. This problem is a combination of dynamic calibration problem, from Ch. 4.3 and 4.4, with indirect state estimation. For focusing only on this issue, ideal sensors are assumed, such that the problem is formulated for the system model from Ch. 4.5.1, given here after taking Laplace transform for zero initial conditions

$$(s \cdot \mathbf{I}_{11} - \mathbf{a}_{11}) \cdot \mathbf{X}_1(s) = \mathbf{a}_{12} \cdot \mathbf{X}_2(s) + \mathbf{b}_1 \cdot \mathbf{U}(s)$$
$$(s \cdot \mathbf{I}_{22} - \mathbf{a}_{22}) \cdot \mathbf{X}_2(s) = \mathbf{a}_{21} \cdot \mathbf{X}_1(s) + \mathbf{b}_2 \cdot \mathbf{U}(s)$$
$$\mathrm{Y}(s) = \mathrm{c}_1 \cdot \mathrm{X}_1(s)$$

The solution of the inverse problem for this LTI ODE system can be obtained analytically for these three equations with three unknowns

$$\mathbf{X}_1(s) = \mathbf{c}_1^{-g} \cdot \mathbf{Y}(s)$$
$$\mathbf{U}_{est}(s) = [\mathbf{a}_{12} \cdot (s \cdot \mathbf{I}_{22} - \mathbf{a}_{22})^{-1} \cdot \mathbf{b}_2 + \mathbf{b}_1]^{-g} \cdot [(s \cdot \mathbf{I}_{11} - \mathbf{a}_{11}) - \mathbf{a}_{12} \cdot (s \cdot \mathbf{I}_{22} - \mathbf{a}_{22})^{-1} \cdot \mathbf{a}_{21}] \cdot \mathbf{c}_1^{-g} \cdot \mathbf{y}(s)$$
$$\mathbf{X}_{2est}(s) = \{[(s \cdot \mathbf{I}_{22} - \mathbf{a}_{22})^{-1} \cdot \mathbf{a}_{21} + (s \cdot \mathbf{I}_{22} - \mathbf{a}_{22})^{-1} \cdot \mathbf{b}_2 \cdot [\mathbf{a}_{12} \cdot (s \cdot \mathbf{I}_{22} - \mathbf{a}_{22})^{-1} \cdot \mathbf{b}_2 + \mathbf{b}_1]^{-g} \cdot [(s \cdot \mathbf{I}_{11} - \mathbf{a}_{11}) - \mathbf{a}_{12} \cdot (s \cdot \mathbf{I}_{22} - \mathbf{a}_{22})^{-1} \cdot \mathrm{a}_{21}]\} \cdot \mathbf{c}_1^{-g} \cdot \mathbf{y}(s)$$

These solutions are computationally intensive even for LTI ODE systems and its accuracy strongly depends on the number and location of sensors producing y(t).

Example 4.17 Scalar equations for a SISO system with two states are

$$(s - a_{11}) \cdot X_1(s) = a_{12} \cdot X_2(s) + b_1 \cdot U(s)$$

$$(s - a_{22}) \cdot X_2(s) = a_{21} \cdot X_1(s) + b_2 \cdot U(s)$$

$$Y(s) = c_1 \cdot X_1(s) + 0 \cdot X_2(s)$$

The solutions for $X_1(s)$, $X_2(s)$ and U given Y are

$$X_1(s) = Y(s) / c_1$$

$$U_{est}(s) = \{[(s - a_{11}) - a_{12} \cdot a_{21} / (s - a_{22})] / [a_{12} \cdot b_2 / (s - a_{22}) + b_1]\} \cdot y(s) / c_1$$

$$X_{2est}(s) = \{[a_{21} / (s - a_{22}) + b_2 \cdot [(s - a_{11}) - a_{12} \cdot a_{21} / (s - a_{22})] / [(s - a_{22}) [a_{12} \cdot b_2 / (s - a_{22}) + b_1]]\} \cdot c_1^{-g} \cdot y(s)$$

Example 4.18 For the under-actuated and under-sensed mechanical system shown in Fig. 4.36, obtain x_1 and F_2 given $y = x_2$.

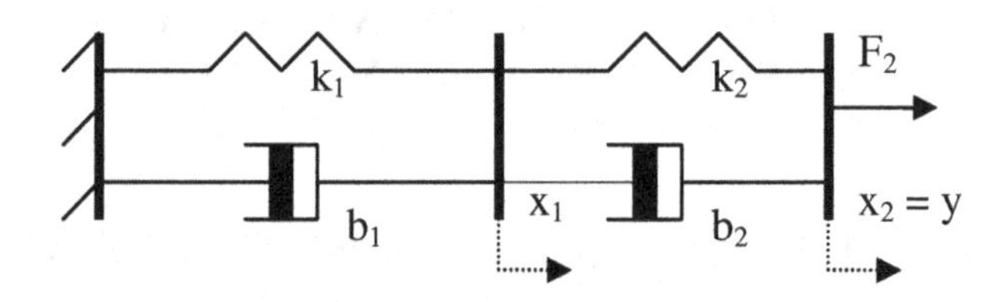

Fig. 4.36 Under-actuated and under-sensed mechanical system

Motion equations are

$$F_2 = k_2 \cdot (x_2 - x_1) + b_2 \cdot (dx_2 / dt - dx_1 / dt)$$
$$k_1 \cdot x_1 + b_1 \cdot dx_1 / dt + k_2 \cdot (x_1 - x_2) + b_2 \cdot (dx_1 / dt - dx_2 / dt) = 0$$

Laplace transform for zero initial conditions give the solutions

$$x_{1,est} = \frac{k_2 + b_2 \cdot s}{k_1 + k_2 + (b_1 + b_2) \cdot s} y(s)$$

$$F_{2,est} = \frac{(k_1 + b_1 \cdot s)(k_2 + b_2 \cdot s)}{k_1 + k_2 + (b_1 + b_2) \cdot s} y(s)$$

Problems

1. Consider a K thermocouple with gain at 20 [^{0}C] of 40 [μV/^{0}C\ and time constant of 0.005 [s].
   a) Obtain the unit step response and the Bode diagram
   b) Calculate the cutoff frequency defining the bandwidth.
   c) Using MATLAB, obtain the harmonic response for a unit amplitude sinusoidal input with ω = 100, 200 and 1000 [rad/s].
   d) Verify that the cutoff frequency defining the bandwidth corresponds to the defined reduction of the amplitude of the harmonic response.
   e) Make a table of the amplitude variation of the steady state response with ω.

2. Consider a second order instrument with k = 2, ω_n = 200 [rad/s].
   a) Obtain MATLAB simulations for unit step response and Bode diagrams for the of the damping ratio values ζ = 0, 0.3, 0.7, 1.0, 2.0.
   b) Calculate analytically the amplitude and phase of the harmonic response for a unit amplitude sinusoidal input with ω = 100, 200 and 1000 [rad/s].
   c) Calculate analytically the cutoff frequency defining the bandwidth.
   d) Obtain MATLAB Bode diagrams of the inverse dynamics compensator for the following values of the damping ratio ζ = 0, 0.3, 0.7, 1.0 and 2.0.
   e) At what frequency the magnitude is ten times higher than at 10 [rad/s]?

3. Consider the mechanical system shown in Fig. 4.36 for $k_1 = k_2 = k$ and $b_1 = b_2 = 0$.

a) Obtain estimations for x_1 and F_2.
b) Are the results intuitive? Could the results be obtained without any calculation?

4. Consider the under-actuated and under-sensed mechanical system shown in Fig. 4.37. Obtain estimations for F_2 and x_2.

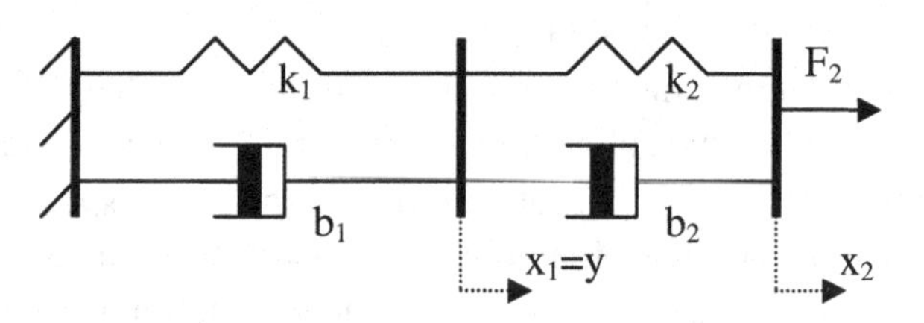

Fig. 4.37 Under-actuated and under-sensed system with measured output $y = x_1$

Chapter 5

Active Vibration Control in Flexible Structures

In this chapter modeling and control of various flexible structure systems (shafts, beams and membranes), will be investigated in view of outlining active vibration control issues within the framework of solutions to direct and inverse problems.

In the first part, SISO (Single Input Single Output) and MIMO (Multiple Input Multiple Output) lumped parameters mechanical models will be used for presenting active vibration control approaches [109]. Direct problems and feedback control for shafts, beams and membranes will be investigated in the subsequent sections. MAPLE and FEMLAB based examples of membrane transversal vibration will be presented in the last section.

5.1 Active Vibration Suppression for Lumped Parameters Mechanical Systems Using Force and Position Control

5.1.1 *Direct Problem*

Vibration suppression is illustrated here for the case of harmonic excitations.

A simple SISO translational lumped parameters system, shown in Fig. 5.1, can be used to illustrate the concept of active vibration suppression [17]. Its free body diagram is shown in Fig. 5.2.

The force input subject to control is f(t), the external force perturbation is $f_{ext}(t)$ and the displacement output is y(t).

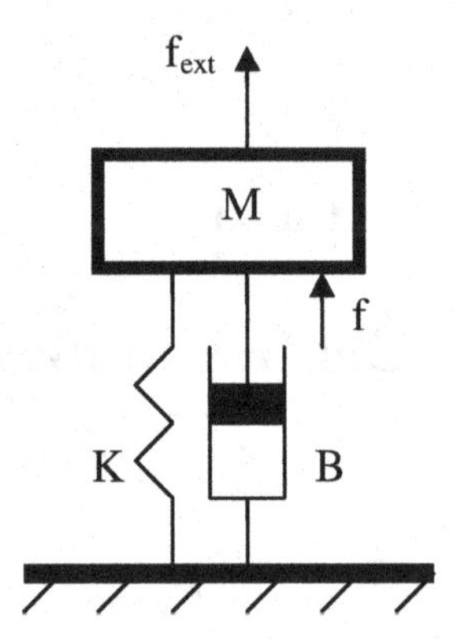

Fig. 5.1 Active vibration suppression in a single input single output system

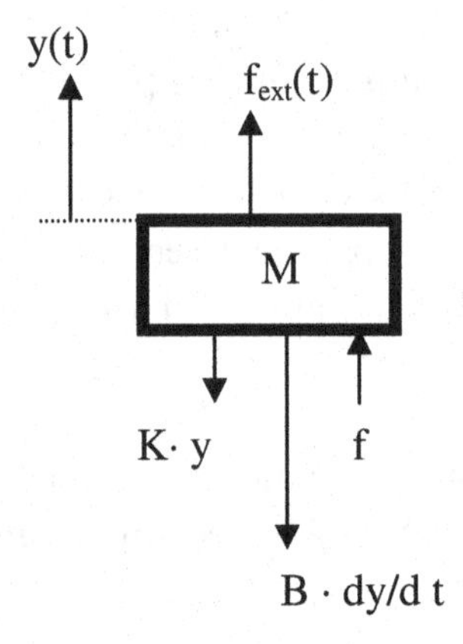

Fig. 5.2 Free body diagram for the system shown in Fig. 5.1

The equation of motion is

$$M \cdot d^2y/dt^2 + B \cdot dy/dt + K \cdot y = f(t) + f_{ext}(t)$$

After taking Laplace transform for zero initial conditions, the result is

$$[Ms^2 + Bs + K] \cdot y(s) = f(s) + f_{ext}(s)$$

The solution gives the direct problem formulation for force input and displacement output

$$y(s) = (f(s) + f_{ext}) / [Ms^2 + Bs + K] =$$
$$f(s) / [Ms^2 + Bs + K] + f_{ext} / [Ms^2 + Bs + K] = y_c(s) + y_a(s)$$

where

$$y_c(s) = f(s) / [Ms^2 + Bs + K]$$

is the displacement due to the controlled force f, and

$$y_a(s) = f_{ext} / [Ms^2 + Bs + K]$$

is the displacement due to the perturbation force f_{ext}.

Perturbation f_{ext} effect cancellation is ideally achieved as

$$y_c(s) + y_a(s) = 0$$

or

$$f(s) + f_{ext}(s) = 0$$

For this LTI system, the perturbation is exemplified by an external harmonic perturbation force

$$f_{ext}(t) = -F_{ext} \cdot \sin \omega t$$

Vibration suppression can be realized by:

- force control that has to achieve

$$f(t) + f_{ext}(t) \rightarrow 0$$

-position control that has to achieve

$$y_c(t) + y_a(t) \rightarrow 0$$

5.1.2 *Force Control for SISO Mechanical System*

Force control approach, shown in Fig. 5.4, is based on:
- force measurement of perturbation $f_{ext}(t)$
- generation of Force Control command $f^{(c)}$ to the actuator to produce an applied controlled force f(t) with the same frequency and amplitude as $f_{ext}(t)$, but of opposed phase, i.e.

$$f(t) = -f_{ext}(t) = -F_{ext} \cdot \sin \omega t = F_{ext} \cdot \sin(\omega t + \pi)$$

This approach is in fact force feedback control for achieving the desired value $f_d = 0$ for the total force applied on M, i.e. $f + f_{ext} \rightarrow 0$, or force regulation, and was proposed for vibration suppression in flexible structures subject to a known or measurable external harmonic excitation, as well as for noise suppression. Force Control command is given by

$$f^{(c)}(t) = -f_{ext}(t)$$

The controller does only a sign change of the input, which is equivalent to a P-control with unity gain. More complex state feedback controllers can also be developed.

This solution requires a force sensor and a force controlled actuator. Force control for vibration suppression implies an implementation with insignificant delays, i.e. ideal sensor $f_{exp} = f_{ext}$, controller $G_c(s) = 1$, and actuator, $G_a(s) = 1$. Ideal force control results in $f(t) = f^{(c)}(t) = -f_{exp}(t) = -f_{ext}(t)$ or

$$f(t) + f_{ext}(t) = 0$$

Control law $G_c(s)$ can be a PD control of force error $f_d - f_{exp}$.

Position control can achieve desired perturbation cancellation using a position sensor and a position controlled actuator.

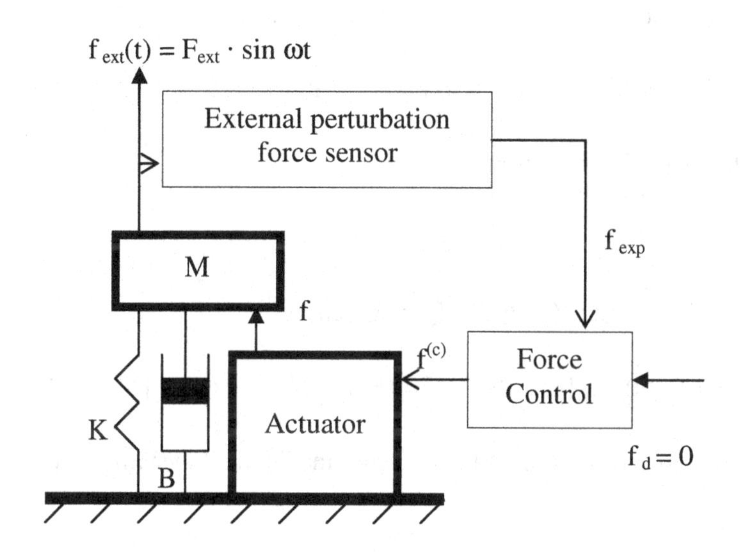

Fig. 5.3 Force control scheme for vibration suppression for a single input single output system subject to an external excitation $f_{ext} = F_{ext} \cdot \sin \omega t$

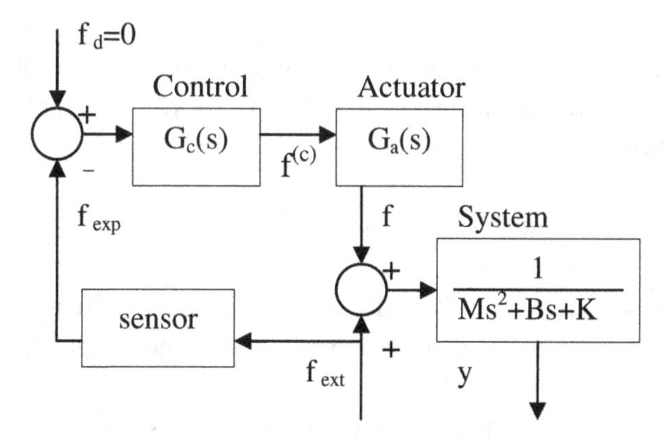

Fig. 5.4 Block diagram for vibration suppression

5.1.3 *Position Feedback Control Approach*

The vibrations of this system are characterized, for $f = 0$, by the displacement $y_a(t)$ due to the perturbation force f_{ext}

$$f_{ext}(t) = F_{ext} \cdot \sin \omega t$$

which has the Laplace transform

$$f_{ext}(s) = F_{ext} \cdot \omega / (s^2 + \omega^2)$$

The displacement $y_a(s)$ due to f_{ext} is given by

$$y_a(s) = f_{ext}(s) / [Ms^2 + Bs + K] = [F_{ext} \cdot \omega / (s^2 + \omega^2)] / [Ms^2 + Bs + K]$$

Control force f(s), applied to the same mechanical system, produces the displacement

$$y_c(s) = f(s) / [Ms^2 + Bs + K]$$

The superposition gives

$$y = y_c + y_a$$

Feedback control of the position has the goal to achieve that y(t) tends towards the desired position $y_d = 0$, i.e. $y(t) \rightarrow y_d$, or, equivalently, to make position error

$$y_d - y = y_d - (y_c(s) + y_a(s)) \rightarrow 0$$

This approach requires the displacement measurement $y_{est}(t)$ and feedback position control that tries to reduce in time the position error

$$y_d - y(t) = - y(t) = - (y_c(t) + y_a(t)) \rightarrow 0$$

The equation of motion for the system shown in Fig. 5.3 is

$$M \cdot d^2y/dt^2 + B \cdot dy/dt + K \cdot y = f(t) + f_{ext}(t)$$

After taking Laplace transform for zero initial conditions, the solution was

$$y(s) = \frac{f(s) - f_{ext}(s)}{M \cdot s^2 - B \cdot s - K}$$

or

$$y(s) = \frac{f(s) + f_{ext}(s)}{K} \cdot \frac{\omega_n^2}{s^2 + 2\varsigma \cdot \omega_n \cdot s + \omega_n^2}$$

where

$\omega_n = \sqrt{(K/M)}$ is the natural frequency
$\zeta = B/(2\sqrt{(K \cdot M)})$ is the damping ratio.

In the under-damped case ($\zeta < 1$) damped natural frequency is $\omega_n \sqrt{(1-\zeta^2)}$.

If $\omega = \omega_n \sqrt{(1-\zeta^2)}$, the system will be in resonance and the amplitude of y(t) will increase significantly. Given that the frequency of the external excitation cannot be changed, active vibration control can be used to change the natural frequency such that the resonance is avoided.

Active vibration reduction using position control is used for:

- vibration isolation to reduce vibrations transmission to and from vibrating bodies
- modification of the mass-spring-damper parameters of the vibrating body.

Active vibration reduction creates an artificial impedance between the vibrating body and the base and, by measuring the displacement and using a desired artificial impedance, generates a force applied to the vibrating body by an actuator. This is achieved by producing f(t), the output of an actuator under PD control command

$$u^{(c)} = - b \cdot dy / dt - k \cdot y$$

where b and k are the PD controller gains and which be interpreted as an artificial b-k impedance.
The effect of the actuator with transfer function

The equation of motion of the controlled system for $f = u^{(c)}$ becomes

$$M \cdot d^2y / dt^2 + B \cdot dy / dt + K \cdot y = - b \cdot dy / dt - k \cdot y + f_{ext}(t)$$

or

$$M \cdot d^2y / dt^2 + (B + b) \cdot dy / dt + (K + k) \cdot y = f_{ext}(t)$$

and the new natural frequency and the damping ratio.

$$\omega_N = \sqrt{((K + k)/M)} > \omega_n$$
$$\zeta_N = (B + b)/(2 \cdot \sqrt{((K + k) \cdot M)})$$

such that the resonance is avoided. Moreover, the coefficient b permits to modify the damping ratio as desired.

The system with active vibration position control subsystem is shown in Fig. 5.5. This control scheme, shown in Fig. 5.5 (b), is implemented using position and velocity sensors and PD control that generates the position command $u^{(c)}$ for the Actuator.

Lumped parameters systems have a finite number of natural frequencies and, in principle, when subject to external harmonic excitations, the amplitude of their vibrations can be reduced using either force control or position control. These control approaches will be applied in the next sections for the control of vibrations various flexible structures.

5.2 Direct Problem and Under-Actuated Control of a Non-Minimum Phase Flexible Shaft

Distributed parameters modeling for flexible structures use often second order differential equations and finite elements models [30]. In this section a simple series system containing a flexible shaft and actuators will be investigated.

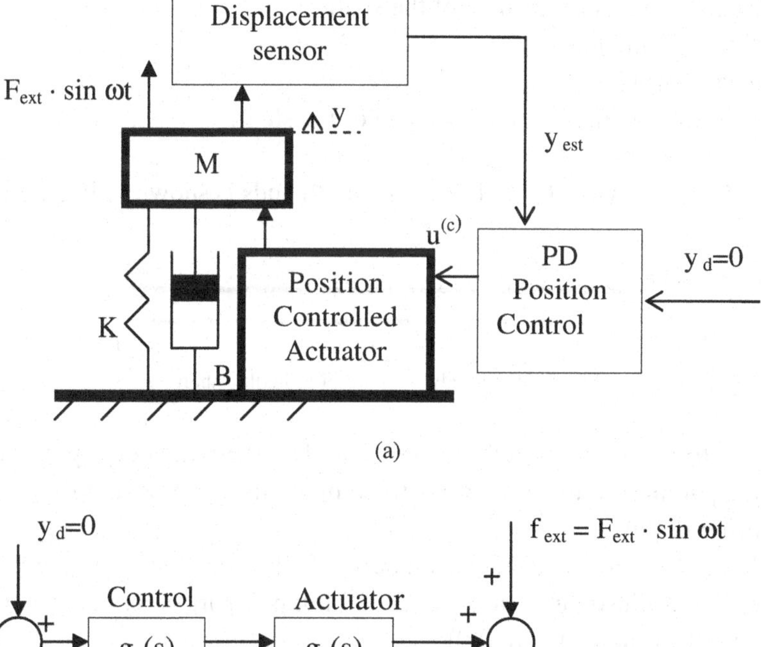

Fig. 5.5 Position control of vibrations scheme (a) and block diagram (b) for a SISO system subject to an external excitation $F_{ext} \sin \omega t$

The distributed parameters model of a constant cross-section and small diameter shaft is given by [17, 30]

$$\frac{\partial^2\theta(x,t)}{\partial t^2} - G/\rho\frac{\partial\theta^2(x,t)}{\partial x^2} = \tau(x,t)\cdot\delta(0)$$

where

θ is the torsional displacement of the shaft
G is the shear modulus
ρ is mass density
τ is the torque applied at the x = 0 end of the shaft.

A flexible shaft with effort-flow cuts at both ends is shown in Fig. 5.6.

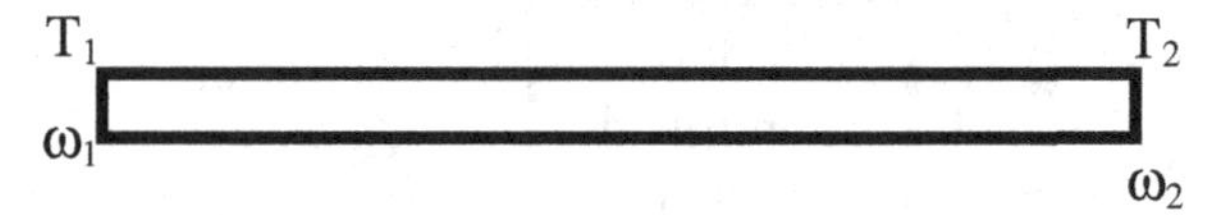

Fig. 5.6 Free body diagram of a flexible shaft

This distributed parameters system can be approximated by various lamped parameters models, easier to compute in real time and to use for controller design.

Three lumped parameters models of the flexible shaft will be presented to illustrate various levels of lumped parameters approximations of a flexible structure [30]:

(a) lumped parameter model with torsional spring coefficient k;
(b) lumped parameter model with torsional spring coefficient k and lumped inertia J;
(c) single finite element model;

These models are obtained as follows:

(a) for the flexible shaft represented by a lumped parameters model with torsional spring coefficient k, the following equations can be obtained:

$$T_1(t) = k(\vartheta_1(t) - \vartheta_2(t))$$
$$T_2(t) = - k ((\vartheta_1(t) - \vartheta_2(t)) = - T_1(t)$$

Given that $\omega = d\vartheta/dt$ and using Laplace transform, this system can be solved to obtain the relationship between the pairs of across-through variables $\{T_1, \omega_1\}$ and $\{T_2, \omega_2\}$

$$T_1(s) = - T_2(s)$$
$$\omega_1(s) = - [1/(k / s)] \cdot T_2(s) + \omega_2(s)$$

or, in matrix form

$$\begin{bmatrix} T_1(s) \\ \omega_1(s) \end{bmatrix} = \begin{bmatrix} -1 & 0 \\ -s/k & 1 \end{bmatrix} \begin{bmatrix} T_2(s) \\ \omega_2(s) \end{bmatrix}$$

This model is suitable only for shafts with low moment of inertia and ignores non-minimum phase property of flexible shafts.

(b) for the flexible shaft represented by a torsional spring coefficient k and lumped inertia J, the following matrix equation can be written for the case of splitting J into two J/2 at the two ends of a spring k:

$$T_1(s) = (J / 2) \cdot s^2 \cdot \vartheta_1 (s) + k \cdot (\vartheta_1(s) - \vartheta_2(s))$$
$$T_2(t) = (J / 2) \cdot s^2 \cdot \vartheta_2(s) + k \cdot (\vartheta_2(s) - \vartheta_1(s))$$

or, in matrix form:

$$\begin{bmatrix} T_1(s) \\ T_2(s) \end{bmatrix} = \left(s^2 \begin{bmatrix} (J/2) & 0 \\ 0 & (J/2) \end{bmatrix} + \begin{bmatrix} k & -k \\ -k & k \end{bmatrix} \right) \cdot \begin{bmatrix} \theta_1(s) \\ \theta_2(s) \end{bmatrix}$$

This equation shows that the flexible shaft is represented by an inertia matrix with no cross-coupling terms and a compliance matrix with cross-couplings.

In case that the right hand side end of the shaft is subject to the torque

$T_1 = \tau_1$ applied by the actuator with a negligible moment of inertia J_1 and right hand side end of the shaft is free, the model becomes

$$\begin{bmatrix} \tau_1(s) \\ 0 \end{bmatrix} = \left(s^2 \begin{bmatrix} (J/2) & 0 \\ 0 & (J/2) \end{bmatrix} + \begin{bmatrix} k & -k \\ -k & k \end{bmatrix} \right) \cdot \begin{bmatrix} \theta_1(s) \\ \theta_2(s) \end{bmatrix}$$

After eliminating $\theta_1(s)$, the following transfer function is obtained [30]

$$\frac{\theta_2(s)}{\tau_1(s)} = \frac{J^2 \cdot s^2 / 2 + k}{s^2 \cdot [J^2 \cdot s^2 / 4 + k \cdot J)]}$$

This shows a minimum phase model, while the flexible shaft is a non-minimum phase system. This model was called inconsistent; a finite element model can be used to obtain a consistent model.

(c) for the flexible shaft represented by a single finite element model, the following equations can be obtained:

$$T_1(s) = (J / 3) \cdot s^2 \cdot \vartheta_1(s) + (J / 6) \cdot s^2 \cdot \vartheta_2(s) + k \cdot (\vartheta_1(s) - \vartheta_2(s))$$
$$T_2(s) = (J / 6) \cdot s^2 \cdot \vartheta_1(s) + (J / 3) \cdot s^2 \cdot \vartheta_2(s) + k \cdot (\vartheta_2(s) - \vartheta_1(s))$$

or in matrix form:

$$\begin{bmatrix} T_1(s) \\ T_2(s) \end{bmatrix} = \left(s^2 \begin{bmatrix} (J/3) & (J/6) \\ (J/6) & (J/3) \end{bmatrix} + \begin{bmatrix} k & -k \\ -k & k \end{bmatrix} \right) \cdot \begin{bmatrix} \theta_1(s) \\ \theta_2(s) \end{bmatrix}$$

This equation shows that the flexible shaft is represented by an inertia matrix and a compliance matrix with cross-coupling.

In case that the left hand side end of the shaft is subject to the torque ($T_1 = \tau_1$) applied by the actuator and right hand side end of the shaft is free ($T_2 = 0$), the model becomes

$$\begin{bmatrix} \tau_1(s) \\ 0 \end{bmatrix} = \left(s^2 \begin{bmatrix} (J/3) & (J/6) \\ (J/6) & (J/3) \end{bmatrix} + \begin{bmatrix} k & -k \\ -k & k \end{bmatrix} \right) \cdot \begin{bmatrix} \theta_1(s) \\ \theta_2(s) \end{bmatrix}$$

Second scalar equation

$$(J \cdot s^2/6 - k) \cdot \theta_1(s) + (J \cdot s^2/3 + k) \cdot \theta_2(s) = 0$$

gives

$$\theta_1(s) = \frac{J \cdot s^2/3 + k}{k - J \cdot s^2/6}\theta_2(s)$$

Eliminating θ_1 (s) from the first scalar equation, the following transfer function is obtained [30]

$$\frac{\theta_2(s)}{\tau_1(s)} = \frac{k - J \cdot s^2/6}{Js^2[Js^2/12 + k)]}$$

This is a non-minimum phase model, i.e. a consistent model of the flexible shaft, while

$$\frac{\theta_1(s)}{\tau_1(s)} = \frac{k + J \cdot s^2/3}{J \cdot s^2 \cdot [J \cdot s^2/12 + k)]}$$

is a minimum phase model.

For the case (c) for $T_1 = \tau_1$ and $T_2 = 0$, a closed loop control to bring θ_1 (s) towards θ_1 (s) is shown in Fig. 5.7. In Fig. 5.7 besides the closed loop control of θ_1 (s) there is open loop dynamics for θ_2 (s), due to under-actuation.

For the case (c) for $T_1 = \tau_1$ and $T_2 = \tau_2$, a closed loop control can be designed for both θ_1 (s) and θ_2(s), but this controller is applicable to the shaft represented by only one finite element.

Any limited number of finite elements limits the model to lower modes of vibrations represented by the model, i.e. leaves higher modes of vibration unaccounted for. Moreover, the closed loop control of the vibrations of a flexible shaft is subject to the limitations of any infinite dimensional system controlled by a finite number of point actuators.

More detailed analysis of active control of flexible structures is presented in the subsequent sections.

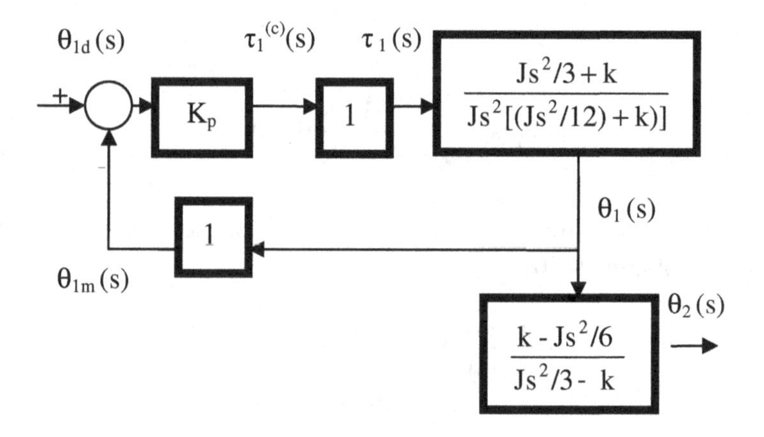

Fig. 5.7 Closed loop control of $\theta_1(s)$

5.3 Control of Vibrations in Beams

5.3.1 *Perturbation Cancellation Control in MIMO Linear Systems*

Before the presentation of the control of vibrations in beams and plates, the general control approach will be presented for the case of a Multi Input Multi Output (MIMO) linear system subject to an external perturbation w(t), shown in Fig. 5.8, as a generalization of the above SISO translational lumped parameters system subject to a harmonic perturbation, presented in Ch. 5.1.

The linear time invariant system is modeled by the following linear ordinary differential equations (ODE) used for modeling lumped parameters systems

$$d\mathbf{X(t)}/dt = \mathbf{A} \cdot \mathbf{X(t)} + \mathbf{B} \cdot \mathbf{u}(t) + \mathbf{G} \cdot \mathbf{w}(t)$$
$$\mathbf{y}(t) = \mathbf{C} \cdot \mathbf{X(t)}$$

where
$\mathbf{X}(t)$ = N_x-vector of states with given initial conditions $\mathbf{x}(0)$
$\mathbf{u}(t)$ = N_u-vector of inputs
$\mathbf{w}(t)$ = N_w-vector of disturbances
$\mathbf{y}(t)$ = N_y-vector of outputs
A, B, G, C, time invariant matrices

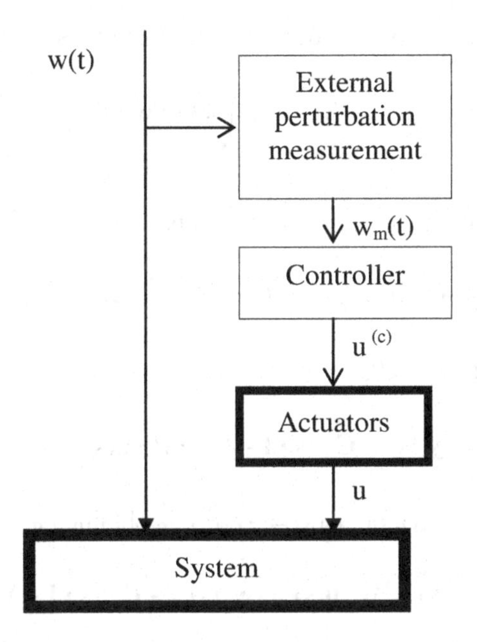

Fig. 5.8 Controller for perturbation effects suppression for a MIMO linear system subject to an external perturbation w(t)

After applying Laplace transform, the model becomes

$$(s \cdot \mathbf{I}\text{-}\mathbf{A}) \cdot \mathbf{X}(s) = \mathbf{B} \cdot \mathbf{u}(s) + \mathbf{G} \cdot \mathbf{w}(s)$$
$$\mathbf{Y}(s) = \mathbf{C} \cdot \mathbf{X}(s)$$

For linear systems, superposition principle can be applied will be analyzed here for the following cases

a) $\mathbf{u}(t) = 0$ and $\mathbf{w}(t) \neq 0$ resulting in the output $\mathbf{y}_w(t)$
b) $\mathbf{u}(t) \neq 0$ and $\mathbf{w}(t) = 0$ for an output $\mathbf{y}_u(t) = -\mathbf{y}_w(t)$
c) superposition of the above cases for the case $\mathbf{u}(t) \neq 0$ and $\mathbf{w}(t) \neq 0$.

a) For $\mathbf{u}(t) = 0$ and $\mathbf{w}(t) \neq 0$
the state and output vectors are denoted $\mathbf{X}_w(s)$ and $\mathbf{y}_w(s)$

$$(s \cdot \mathbf{I} - \mathbf{A}) \cdot \mathbf{X}_w(s) = \mathbf{G} \cdot \mathbf{w}(s)$$
$$\mathbf{y}_w(s) = \mathbf{C} \cdot \mathbf{X}_w(s)$$

such that the output due to the perturbation is

$$\mathbf{y}_w(s) = \mathbf{C} \cdot (s \cdot \mathbf{I} - \mathbf{A})^{-1} \cdot \mathbf{G} \cdot \mathbf{w}(s)$$

b) For $\mathbf{u}(t) \neq 0$ and $\mathbf{w}(t) = 0$,

$$\mathbf{y}_u(s) = \mathbf{C} \cdot (s \cdot \mathbf{I} - \mathbf{A})^{-1} \cdot \mathbf{B} \cdot \mathbf{u}(s)$$

The output $\mathbf{y}_u(t) = -\mathbf{y}_w(t)$, or $\mathbf{y}_u(s) = -\mathbf{y}_w(s)$, is obtained as

$$\mathbf{y}_u(s) = \mathbf{C} \cdot (s \cdot \mathbf{I} - \mathbf{A})^{-1} \cdot \mathbf{B} \cdot \mathbf{u}(s) = -\mathbf{y}_w(s) = -\mathbf{C} \cdot (s \cdot \mathbf{I} - \mathbf{A})^{-1} \cdot \mathbf{G} \cdot \mathbf{w}(s)$$

The condition for perturbation effect cancellation is

$$\mathbf{B} \cdot \mathbf{u}(s) = -\mathbf{G} \cdot \mathbf{w}(s)$$

For achieving

$$\mathbf{y}_u(t) + \mathbf{y}_w(t) \rightarrow 0$$

requires that

$$\mathbf{u}(s) = -\mathbf{B}^{-1} \cdot \mathbf{G} \cdot \mathbf{w}(s)$$

such that, after inverse Laplace transform, $\mathbf{u}^{(c)}(t)$, the command for the feedfback controller is

$$\mathbf{u}^{(c)}(s) = -\mathbf{B}^{-1} \cdot \mathbf{G} \cdot \mathbf{w}(s)$$

If the matrix $\mathbf{B}$ $[N_u \cdot N_x]$ is not an invertible matrix the generalized inverse has to be used [19].

c) The superposition of the above cases for the case $\mathbf{u}(t) \neq 0$ and $\mathbf{w}(t) \neq 0$, results in an overall output

$$\mathbf{y}(s) = \mathbf{y}_u(s) + \mathbf{y}_w(s) = 0$$

or, after inverse Laplace transform

$$\mathbf{y}(t) = \mathbf{y}_u(t) + \mathbf{y}_w(t) = 0$$

showing the condition for the perturbation effect suppression.

This ideal result of perturbation effect cancellation using a feedback controller with input $\mathbf{w}(t)$, output $\mathbf{u}^{(c)}(t)$ and the gain $-\mathbf{B}^{-1}\mathbf{G}$ is conditioned by the numerous assumptions made: linear time invariant system with invertible matrix $\mathbf{B}$, *i.e.* with as many inputs as states, with perfectly measurable perturbations ($\mathbf{w}(t) = \mathbf{w}_m(t)$), with actuators that produce outputs exactly as the commands are ($\mathbf{u}(t) = \mathbf{u}^{(c)}(t)$) etc. These assumptions are generally not valid for practical systems and instead of

$$\mathbf{y}_u(t) + \mathbf{y}_w(t) = 0$$

feedback control can be made to achieve

$$\mathbf{y}_u(t) + \mathbf{y}_w(t) \rightarrow 0$$

A better transient regime can be achieved by PD control, as shown in Ch. 5.1 for SISO systems.

5.3.2 *Direct Problem in Beam Vibration Modeling*

The analysis of a beam vibrations requires the use if the solution y(x, t) of the Euler-Bernoulli beam equation

$$a^2 \frac{\partial^4 y}{\partial x^4} + \frac{\partial^2 y}{\partial t^2} = \frac{F(x,t)}{\rho \cdot A \cdot L} = \frac{f(x,t)}{\rho \cdot A}$$

where

$$a = \sqrt{\frac{E \cdot I}{\rho \cdot A}}$$

in $[m^2/s]$ is beam coefficient, F(x, t) in [N] is the applied force, and f(x, t) in [N/m] is the force per unit length.

The method of separation of variables assumes a solution in the form [17]

$$y(x, t) = X(x) \cdot T(t)$$

The free vibration of the beam corresponds to f(x, t) = 0, such that

$$a^2 \frac{\partial^4 y(x,t)}{\partial x^4} + \frac{\partial^2 y(x,t)}{\partial t^2} = 0$$

Applying the above assumed solution, beam equation becomes

$$a^2 \frac{d^4 X(x)T(t)}{dx^4} + \frac{d^2 X(x)T(t)}{dt^2} = 0$$

or

$$a^2 \frac{X^{iv}(x)}{X(x)} = -\frac{T''(t)}{T(t)} = \omega^2$$

The meaning of the α^2 will be clarified later.

The above equalities can be separated in two ordinary differential equations, one second order temporal equation

$$T''(t) + \omega^2 T(t) = 0$$

and a second one, fourth order spatial equation

$$X^{iv}(x) - \frac{\omega^2}{a^2} X(x) = 0$$

where

$$a = \sqrt{\frac{EI}{\rho A}}$$

$$\frac{\omega^2}{a^2} = \frac{\omega^2 \rho A}{EI} = \beta^4$$

and the wave number in [1/m] is

$$\beta = \sqrt{\frac{\omega}{a}}$$

The general solution of the temporal equation is

$$T(t) = A \cdot \cos(\omega t) + B \cdot \sin(\omega t)$$

where the constants A and B are determined by the required two initial conditions of T(t) and dT(t) / dt for t = 0. For the above equation is obvious now that the assumed constant α^2 corresponds to the natural frequency of oscillation $\omega = 2 \cdot \pi \cdot f$.

The general solution of the fourth order spatial equation is based on the assumed general solution

$$X(x) = a \cdot e^{-j\beta x} + b \cdot e^{j\beta x} + c \cdot e^{-\beta x} + d \cdot e^{\beta x}$$

Given Euler identities

$$e^{\pm\beta x} = \cosh\beta \cdot x \pm \sinh\beta \cdot x$$

$$e^{\pm\beta \cdot x} = \cos\beta \cdot x \pm \sin\beta \cdot x$$

the above solution can be rewritten in a form that identifies the mode shapes of the beam

$$X(x) = C \cdot \cos(\beta x) - D \cdot \sin(\beta x) - E \cdot \cosh(\beta x) - F \cdot \sinh(\beta x)$$

where C, D, E, F can be obtained as function of a, b, c, d. Actual values for C, D, E, F result from the required four boundary conditions for the fourth order spatial equation of the beam.

Appendix A presents the solution of Euler-Bernoulli beam equation using the method of separation of variables for the transversal forced vibrations of a cantilever beam. The beam is subject to a continuous sinusoidal excitation at point $x = l_p$

$$f(x,t) = \rho \cdot A \cdot \alpha \cdot \sin(\Omega in \cdot \delta(x - l_p)$$

The complete solution for the transversal vibration of a cantilever beam, subject to a single frequency of excitation Ω, is given by [17, 23, 55]

$$y(x,t) = \sum_{n=1}^{\infty} \frac{\alpha X_n(x) X_n(l_p)}{\omega_n^2 - \Omega^2} (\sin\Omega s - \frac{\Omega}{\omega_n} \sin\omega_n t)$$

where $X_n(l_p)$ is the value at $x = l_p$ of the mode shape function $X_n(x)$, given by

$$X_n(l_p) = A_n \left[\cosh\beta_n l_p - \cos\beta_n l_p - \sigma_n(\sinh\beta_n l_p - \sin\beta_n l_p)\right]$$

and

$$\begin{aligned} A_n^2 = {} & 4\beta_n / \{4\beta_n l + 2\sigma_n\cos(2\beta_n l) - 2\sigma_n\cosh(2\beta_n l) - 4\cosh(\beta_n l)\sin(\beta_n l) \\ & - 4\sigma_n^2\cosh(\beta_n l)\sin(\beta_n l) + \sin(2\beta_n l) - \sigma_n^2\sin(2\beta_n l) - 4\cos(\beta_n l)\sinh(\beta_n l) \\ & + 4\sigma_n^2\cos(\beta_n l)\sinh(\beta_n l) + 8\sigma_n\sin(\beta_n l)\sinh(\beta_n l) + \sinh(2\beta_n l) \\ & + \sigma_n^2\sinh(2\beta_n l)\} \end{aligned}$$

The equation for $X_n(l_p)$ for the beam quantifies the effect on mode n of an actuator located at $x = l_p$ on the y(x, t).

The direct problem equation for the output y(x, t) is the model for transversal forced vibrations due to the harmonic excitation of given frequency Ω from input from a point actuator in located at $x = l_p$

$$f(x,t) = \rho \cdot A \cdot \alpha \cdot \sin(\Omega in \cdot \delta(x - l_p)$$

This solution for single frequency excitation cannot be used for designing feedback control for a beam, because the applied force per unit length, f(x, t), is too restrictively assumed of a single frequency and applied at a boundary point $x = l_p$. A generic applied force per unit length, f(x, t), can have an arbitrary time variation and could be distributed along x [35]

$$f(x,t) = \sum_{n=1}^{\infty} P_n \cdot \sin(n \cdot \pi \cdot x/l) \sum_{r=1}^{\infty} P_r \cdot \sin(\Omega_r \cdot t)$$

where l is the length of the beam.

A generic solution for simply supported Euler-Bernoulli beam equation for the initial conditions T(0)=0 and T'(0)=0 and subject to an arbitrary f(x, t), is [17, 23, 55]

$$y(x,t) = \sum_{r=1}^{\infty} \sum_{n=1}^{\infty} \left[\frac{P_n/\overline{m}}{\omega_n^2 - \Omega_r^2}(\sin\Omega_r \cdot t - \frac{\Omega_r}{\omega_n}\sin\omega_n \cdot t)\right] \cdot \sin(n \cdot \pi \cdot x/l)$$

where $\overline{m} = M / l$ is the mass per unit length.

This generic solution for simply supported Euler-Bernoulli beam equation is a particular case of the solution for simply supported plate subject to a distributed force with arbitrary time variation.

The above generic solution to the direct problem is a double infinite series, *i.e.* is not in closed form, and gives the transversal displacement $y(x, t)$ due to the excitation $f(x, t)$. The spatial distribution

$$\sum_{n=1}^{\infty} P_n \cdot \sin(n \cdot \pi \cdot x / l)$$

is assumed based on the n modes of vibration of the beam. Further analysis of the effect of distributed applied force on beam vibration is beyond the scope of this book.

The response to multiple harmonic excitations of various frequencies generated by several actuators located in various points along the beam represents a direct problem of a higher complexity. The output equation for $y(x, t)$ is in the non-closed form of an infinite series of modes and, inverse problem does not have an analytical solution such inverse problem solution is not obtainable analytically. Active control of vibrations has to be based on a solution that does not come from the inverse problem. In fact the solution analyzed in the next section is based on feedback modal control for a reduced number of modes. Moreover, a real beam is not subject to only transversal vibrations across the width of the beam, but also across the thickness of the beam, as well as longitudinal vibrations and torsional vibrations.

5.3.3 *Feedback Control of Transversal Vibrations in Beams*

Figure 5.9 shows the diagram of feedback control for perturbation effects suppression for an infinite beam, based on the generic diagram from Fig. 5.8, for the case of using acceleration for perturbation measurement, actuator output and error $E(x_d, t)$.

Feedback control of transversal vibrations in beams cannot be based on the inverse of a direct problem with infinite dimensional solution of the partial differential equations. A practical approach for the feedback control for a beam, like the one shown in Fig. 5.9, has to be based on reduced order models.

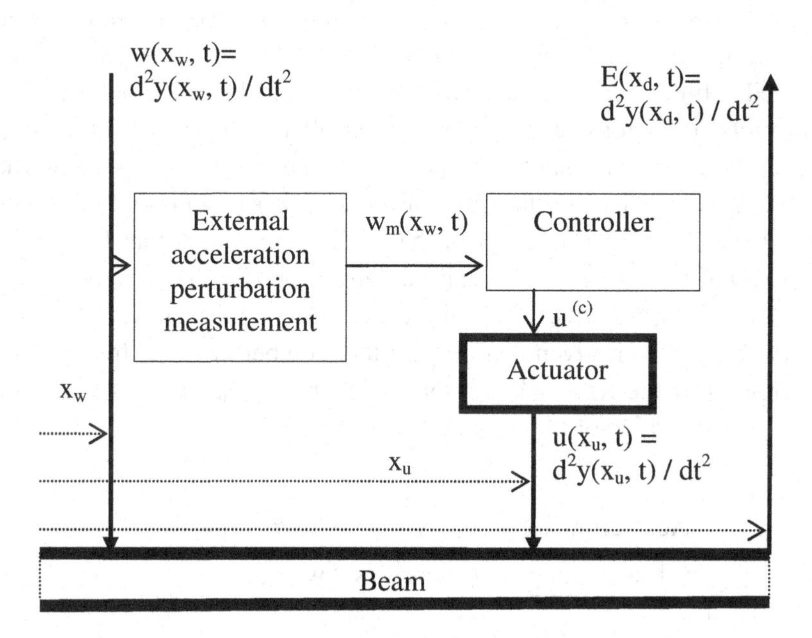

Fig. 5.9 Feedback control for perturbation effects suppression for an infinite beam

One such finite dimensional model is the finite elements model, presented in Ch. 3.3 for a string, that can be reformulated for a beam and the conclusions for the string hold also for the beam, in particular that any limited number of finite elements limits the model to lower modes of vibrations represented by the model, *i.e.* leaves higher modes of vibration unaccounted for.

Another lower order model that can be used for feedback control of a beam, shown in Fig. 5.10, is based on transfer functions, [23].

This model can be used if the beam can be assumed a linear time invariant system.

Transfer functions are unidirectional models defined for a given input and a given output. In a beam, however, waves propagate in both directions and, for this reason, in Fig. 5.10 are shown separate transfer functions designated for each unidirectional propagation case: B(s) from the perturbation $w(x_d, t)$ point x_d, to the error $E(x_d, t)$ point x_d, $B_1(s)$ from perturbation $w(x_d, t)$ point x_d, to the to the actuator location x_u and $B_2(s)$ from perturbation $w(x_d, t)$ to the error $E(x_d, t)$ point x_d. In the assumed infinite beam (or anechoically terminated beam) there are no end points reflections. The feedback controller is defined by the K(s) transfer function, that has to be designed. The effect of this feedback controller materializes in the value of $d_2(x_d, t)$ that is supposed to cancel $d(x_d, t)$ due to the perturbation in the point $x = x_w$ such that $E(x_d, t) = d_2(x_d, t) + d(x_d, t) = 0$. *i.e.* ideally a zero error $E(x_d, t)$. In reality, the cancellation is not perfect and only a low value significant error can be expected. Moreover, given that only a finite number of lower frequencies are targeted by the feedback controller, the error $E(x_d, s)$, for $s = \omega$, will increase as the value of the frequency ω increases.

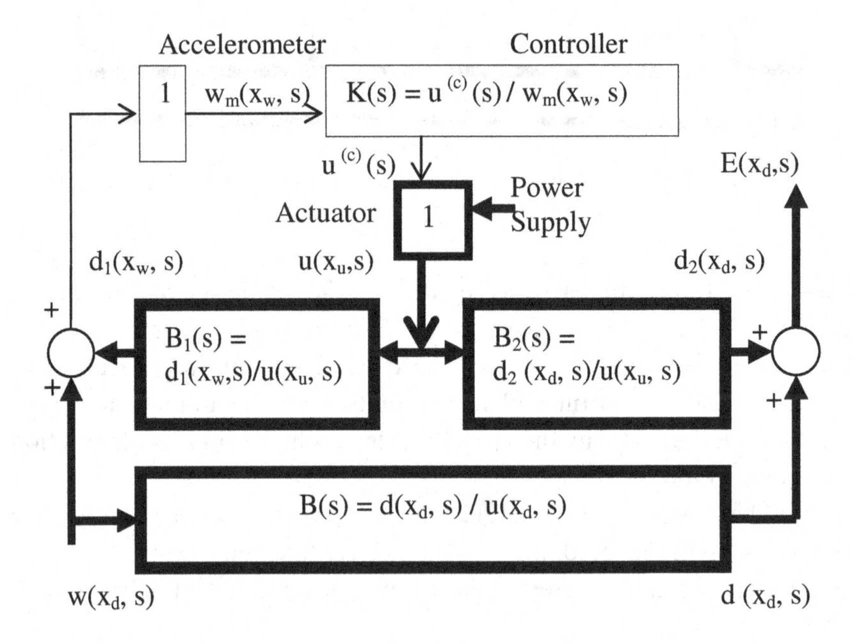

Fig. 5.10 Block diagram of non-collocated feedback control for perturbation effects suppression for a beam using transfer functions

The block diagram from Fig. 5.10 permits to write the following model of the beam feedback control

$$w\,B(s) = d$$
$$u\,B_2(s) = d_2$$
$$u\,B_1(s) = d_1$$
$$d + d_2 = E$$
$$u = K(s) \cdot (w + d_1)$$

From these equations can be obtained the transfer function E(s)/w(s) that characterizes the controller effect in reducing the effect of the perturbation w(s) on the error E(s) [23]. This transfer function can be obtained from the above five equations, by eliminating d, d_1, d_2 and u

$$E(s)/w(s) = B(s) + B_2(s) \cdot K(s) \,/\, [1 - B_1(s) \cdot K(s)]$$

The ideal effect is E(s) = 0 and this requires

$$B(s) + B_2(s)\,K(s) \,/\, [1 - B_1(s) \cdot K(s)] = 0$$

This equation permits to calculate the transfer function of the feedback control

$$K(s) = B(s) \,/\, [B_1(s) \cdot B(s) - B_2(s)]$$

In practical applications, the three transfer function for the above beam model are obtained experimentally using frequency response measurements, [19, 34]. A block diagram for these experiments is shown in Fig. 5.11.

The sinusoidal signal generator receives the command signal $I^{(c)}(\omega) = I^{(c)} \cdot \sin(\omega t)$ over the range of frequencies significant for the beam vibration control, for example 0.01 to 100 Hz., i.e. four orders of magnitude of frequency variation. This input to the sinusoidal signal generator is usually an analog voltage signal and the electromechanical signal generator will have the output $I(\omega) = I \cdot \sin(\omega t)$, a sinusoidal transversal excitation applied to the beam, often with a phase difference ignored here. The output transversal displacement of the beam $O(\omega)$, is measured by a signal transducer that produces the analog voltage $O_m(\omega)$ transmitted to the Analog Input of a Data Acquisition Board installed in a PC.

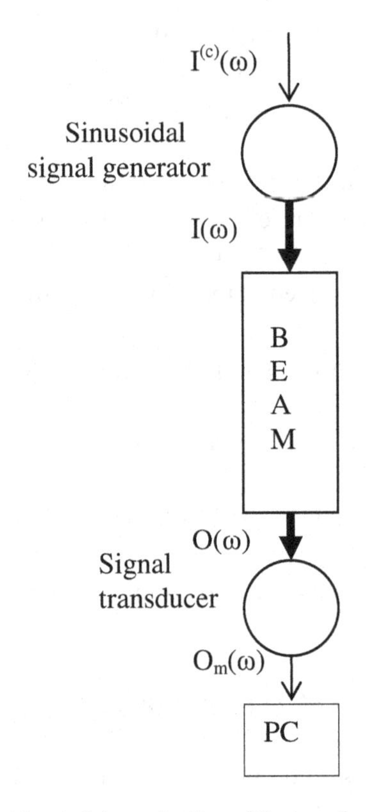

Fig. 5.11 Experimental determination of the transfer function of a beam

A suitable data acquisition and processing software, for example LabVIEW™ or MATLAB™ can be used for obtaining the experimental Bode diagrams amplitude and phase versus frequency ω. Next step is to obtain an approximation of the transfer function of the beam using the experimental Bode diagrams. For example, Log-magnitude diagram can be used to determine asymptotic log-magnitude straight line approximations with slopes restricted to integer multiples of +/- 20 dB/decade [34]. After determining corner frequencies, the damping ratio result from the amount of local resonant peak. The phase diagram is used to check the resulting transfer function obtained from log-magnitude curves. Non-minimum phase behavior, typical for flexible beams, can be identified from the high frequency phase that results 180^0 away in computed phase versus experimental phase diagram, indicating that a positive zero was missing and has to be included in the transfer function. Propagation time of the signals in the beam can be identified from the constant rate of change T between computed and experimental phase angles, i.e $-T\omega$, then the propagation delay is given by the multiplicative factor $e^{-T\omega}$.

An example of experimentally determined transfer function, limited to the first three modes of vibration, could be

$$O_m(\omega) / I^{(c)}(\omega) = k \cdot (s - z) \cdot e^{-T\omega} / [(s + p_1) \cdot (s + p_2) \cdot (s + p_3)]$$

This transfer function includes not only the beam, but also the transfer functions of the sinusoidal signal generator and the of transducer. Assuming that the effect of the transfer functions of the sinusoidal signal generator and the transducer are not significant, the three transfer functions from Fig. 5.10 can be written as

$$B(s) = k \cdot (s - z) \cdot e^{-T\omega} /[(s + p_1) \cdot (s + p_2) \cdot (s + p_3)]$$
$$B_1(s) = k_1 \cdot (s - z_1) \cdot e^{-T_1\omega} /[(s + p^{(1)}_1) \cdot (s + p^{(1)}_2) \cdot (s + p^{(1)}_3)]$$
$$B_2(s) = k_2 \cdot (s - z_2) \cdot e^{-T_2\omega} /[(s + p^{(2)}_1) \cdot (s + p^{(2)}_2) \cdot (s + p^{(2)}_3)]$$

The resulting transfer function of the feedback controller would be in this case:

$$K(s) = B(s) / [B_1(s) \cdot B(s) - B_2(s)] =$$
$$\{k \cdot (s - z) \cdot e^{-T\omega} /[(s + p_1) \cdot (s + p_2) \cdot (s + p_3)]\}/\{k_1 \cdot (s - z_1) \cdot e^{-T_1\omega} \cdot k \cdot$$
$$(s - z) \cdot e^{-T\omega} /[(s + p_1) \cdot (s + p_2) \cdot (s + p_3) \cdot (s + p^{(1)}_1) \cdot (s + p^{(1)}_2) \cdot$$
$$(s + p^{(1)}_3)] - k_2 \cdot (s - z_2) \cdot e^{-T_2\omega} /[(s + p^{(2)}_1) \cdot (s + p^{(2)}_2) \cdot (s + p^{(2)}_3)]\} =$$
$$\{k \cdot (s - z) \cdot e^{-T\omega} / [(s + p_1) \cdot (s + p_2) \cdot (s + p_3)]\}/ \{k \cdot k_1 (s - z) \cdot (s- z_1) \cdot$$

$$e^{-(T+T_1)\omega} / [(s + p_1) \cdot (s + p_2) \cdot (s + p_3) \cdot (s + p^{(1)}_1) \cdot (s + p^{(1)}_2) \cdot (s + p^{(1)}_3)] - k_2 \cdot (s - z_2) \cdot e^{-T_2\omega} /[(s + p^{(2)}_1) \cdot (s + p^{(2)}_2) \cdot (s + p^{(2)}_3)]\}$$

The common denominator of this transfer function is

$$\{k \cdot k_1 \cdot (s - z) \cdot (s - z_1) \cdot e^{-(T+T_1)\omega} \cdot [(s + p^{(2)}_1) \cdot (s + p^{(2)}_2) \cdot (s + p^{(2)}_3)] - k_2 \cdot (s - z_2) \cdot e^{-T_2\omega} \cdot [(s + p_1) \cdot (s + p_2) \cdot (s + p_3) \cdot (s + p^{(1)}_1) \cdot (s + p^{(1)}_2) \cdot (s + p^{(1)}_3)]\}$$

Even when all delays are insignificant, $T = T_1 = T_2 = 0$, the denominator of K(s) is a 7-th order polynomial in s, difficult to implement. Results were reported for third order controllers, but this would limit feedback control to the first three lower frequencies of the transfer function, leaving all other higher frequencies with no control [23]. Significant work has to be done in this case to verify if spillover effects to higher frequencies and other types of vibrations do not lead to highly oscillating or unstable open loop dynamics. Moreover, this transfer function based control is effective only if the parameters do not vary in time.

Feedback modal control is presented in the next section.

5.3.4 *Feedback Modal Control*

Feeedback control of vibrations in beams can be achieved using various methods, as for example, modal control and wave reflection control. In this section modal control will be presented modal control of beam vibration [109].

a) Modal control of a 2 DOF mechanical system

To illustrate the concept of modal control, first will be presented the simple case of an under-actuated 2 DOF mechanical system, shown in Fig. 5.12.

In Ch. 3.2 was derived the model for this system

$$m_1 \cdot d^2 x_1/dt^2 + k_1 \cdot x_1 + k_2 \cdot (x_1 - x_2) = f_1$$

$$m_2 \cdot d^2 x_2/dt^2 + k_2 (x_2 - x_1) = 0$$

with one measured output

$$y_1 = x_1$$

and one input f_1.

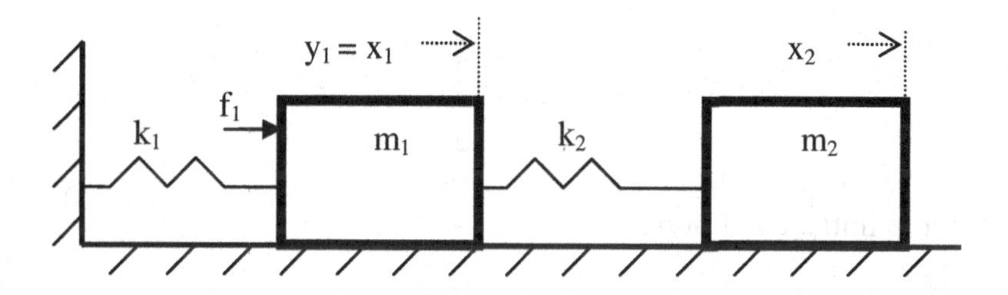

Fig. 5.12 An under-actuated 2 DOF mechanical system

Initial conditions are assumed

$$x_1(0) = x_{10}$$
$$x_2(0) = x_{20}$$
$$dx_1(t) / dt|_{t=0} = v_{10}$$
$$dx_2(t) / dt|_{t=0} = v_{20}$$

In matrix form this equation becomes

$$\mathbf{M} \cdot \ddot{\mathbf{X}} + \mathbf{K} \cdot \mathbf{X} = \mathbf{F}$$

where

$$\mathbf{X} = \begin{bmatrix} x_1 \\ x_2 \end{bmatrix}$$

$$\mathbf{K} = \begin{bmatrix} k_1 + k_2 & -k_2 \\ -k_2 & k_2 \end{bmatrix}$$

$$\mathbf{M} = \begin{bmatrix} m_1 & 0 \\ 0 & m_2 \end{bmatrix}$$

$$\mathbf{F} = \begin{bmatrix} f_1 \\ 0 \end{bmatrix}$$

subject to initial conditions:

$$\mathbf{X}(0) = \begin{bmatrix} x_{10} \\ x_{20} \end{bmatrix}$$

$$\dot{\mathbf{X}}(0) = \begin{bmatrix} v_{10} \\ v_{20} \end{bmatrix}$$

Modal analysis requires to determine the eigenvalues and the eigenvetors of this system and this is facilitated by a first transformation of variables **X** into **Q** [23] using

$$\mathbf{X} = \mathbf{M}^{-1/2} \cdot \mathbf{Q}$$

or

$$\mathbf{Q} = \mathbf{M}^{1/2} \cdot \mathbf{X}$$

where

$$\mathbf{Q} = \begin{bmatrix} q_1 \\ q_2 \end{bmatrix}$$

The above matrix equation of motion becomes

$$\mathbf{M} \cdot \mathbf{M}^{-1/2} \cdot \ddot{\mathbf{Q}} + \mathbf{K} \cdot \mathbf{M}^{-1/2} \cdot \mathbf{Q} = \mathbf{F}$$

Pre-multiplying with $\mathbf{M}^{-1/2}$ gives

$$\mathbf{M}^{-1/2} \cdot \mathbf{M} \cdot \mathbf{M}^{-1/2} \ddot{\mathbf{Q}} + \mathbf{M}^{-1/2} \mathbf{K} \cdot \mathbf{M}^{-1/2} \cdot \mathbf{Q} = \mathbf{M}^{-1/2} \cdot \mathbf{F}$$

where

$$\mathbf{M}^{-1/2} \cdot \mathbf{M} \cdot \mathbf{M}^{-1/2} = \mathbf{I}$$

Denoting

$$\mathbf{M}^{-1/2} \cdot \mathbf{K} \cdot \mathbf{M}^{-1/2} = \tilde{\mathbf{K}}$$

$$\mathbf{M}^{-1/2} \cdot \mathbf{F} = \tilde{\mathbf{F}} = \begin{bmatrix} m_1^{-1/2} & 0 \\ 0 & m_2^{-1/2} \end{bmatrix} \cdot \begin{bmatrix} f_1 \\ 0 \end{bmatrix} = \begin{bmatrix} m_1^{-1/2} \cdot f_1 \\ 0 \end{bmatrix}$$

previous matrix equation of motion becomes

$$\mathbf{I} \cdot \ddot{\mathbf{Q}} + \tilde{\mathbf{K}} \cdot \mathbf{Q} = \tilde{\mathbf{F}}$$

with initial conditions

$$\mathbf{Q}(0) = \begin{bmatrix} \mathrm{M}^{1/2} \cdot x_{10} \\ \mathrm{M}^{1/2} \cdot x_{20} \end{bmatrix}$$

$$\dot{\mathbf{Q}}(0)=\begin{bmatrix}\mathbf{M}^{1/2}\cdot v_{10}\\ \mathbf{M}^{1/2}\cdot v_{20}\end{bmatrix}$$

$\tilde{\mathbf{K}}$ is a symmetric matrix that has the same structure as the definition of squared natural frequency for a single degree of freedom mass-spring-damper system $k/m = k\cdot m^{-1} = m^{-1/2}\cdot k\cdot m^{-1/2}$.

Assuming the solution

$$\mathbf{Q}=\mathbf{V}\cdot e^{j\omega t}$$

such that

$$\ddot{\mathbf{Q}}=-\omega^2\cdot\mathbf{V}\cdot e^{j\omega t}$$

For $\mathbf{F} = 0$, the last matrix equation becomes

$$-\omega^2\cdot\mathbf{V}\cdot e^{j\omega t}+\tilde{\mathbf{K}}\cdot\mathbf{V}\cdot e^{j\omega t}=0$$

or

$$\omega^2\cdot\mathbf{V}=\tilde{\mathbf{K}}\cdot\mathbf{V}$$

or, after denoting

$$\omega^2=\lambda$$

$$\lambda\cdot\mathbf{V}=\tilde{\mathbf{K}}\cdot\mathbf{V}$$

or

$$(\tilde{\mathbf{K}}-\lambda\cdot\mathbf{I})\cdot\mathbf{V}=0$$

The determinant equal to zero

$$\left|(\tilde{\mathbf{K}}-\lambda\cdot\mathbf{I})\cdot\mathbf{V}\right|=0$$

gives the characteristic equation for obtaining eigenvalues.

The solutions λ_1 and λ_2 can be written as a diagonal matrix

$$\mathbf{\Lambda}=\begin{bmatrix}\lambda_1 & 0\\ 0 & \lambda_2\end{bmatrix}$$

known as the matrix of eigenvalues while **V** is the corresponding eigenvector, *i.e.*

$$\mathbf{V}_1=\begin{bmatrix}v_{11}\\ v_{12}\end{bmatrix}\ \text{for}\ \lambda_1$$

and

$$\mathbf{V}_1=\begin{bmatrix}v_{21}\\ v_{22}\end{bmatrix}\ \text{for}\ \lambda_2$$

These eigenvectors correspond to mode shapes.

A second transformation of variables **Q** into **R** is given by [23]

$$\mathbf{Q}=\mathbf{P}\cdot\mathbf{R}$$

such that

$$\mathbf{X}=\mathbf{M}^{-1/2}\cdot\mathbf{Q}=\mathbf{M}^{-1/2}\cdot\mathbf{P}\cdot\mathbf{R}$$

where

$$\mathbf{R}=\begin{bmatrix}r_1\\ r_2\end{bmatrix}$$

and

$$\mathbf{P} = [\mathrm{V}_1 \quad \mathrm{V}_2] = \begin{bmatrix} \mathrm{v}_{11} & \mathrm{v}_{21} \\ \mathrm{v}_{12} & \mathrm{v}_{22} \end{bmatrix}$$

P is the modal matrix, that has the property $\mathbf{P}^{\mathrm{T}} \cdot \mathbf{P} = \mathbf{I}$.

Substituting **Q** by **R** in

$$\mathbf{I} \cdot \ddot{\mathbf{Q}} + \tilde{\mathbf{K}} \cdot \mathbf{Q} = \tilde{\mathbf{F}}$$

gives

$$\mathbf{P} \cdot \ddot{\mathbf{R}} + \tilde{\mathbf{K}} \cdot \mathbf{P} \cdot \mathbf{R} = \tilde{\mathbf{F}}$$

Pre-multiplying by $\mathbf{P}^{\mathrm{T}}$ gives

$$\mathbf{P}^{\mathbf{T}} \cdot \mathbf{P} \cdot \ddot{\mathbf{R}} + \cdot \mathbf{P}^{\mathbf{T}} \tilde{\mathbf{K}} \cdot \mathbf{P} \cdot \mathbf{R} = \mathbf{P}^{\mathbf{T}} \cdot \tilde{\mathbf{\Phi}}$$

Taking into account that $\mathbf{P}^{\mathrm{T}} \cdot \mathbf{P} = \mathbf{I}$ and that $\mathbf{P}^{\mathrm{T}} \cdot \tilde{\mathbf{K}} \cdot \mathbf{P}$ gives the diagonal eigenvalues matrix $\mathbf{\Lambda}$

$$\mathbf{P}^{\mathrm{T}} \cdot \tilde{\mathbf{K}} \cdot \mathbf{P} = \mathbf{\Lambda} = \begin{bmatrix} \lambda_1 & 0 \\ 0 & \lambda_2 \end{bmatrix}$$

the above second order differential equation in

$$\mathbf{R} = \mathbf{P}^{-1} \cdot \mathbf{M}^{1/2} \cdot \mathbf{X} = \mathbf{P}^{\mathrm{T}} \cdot \mathbf{M}^{1/2} \cdot \mathbf{X}$$

becomes

$$\mathbf{I} \cdot \ddot{\mathbf{R}} + \mathbf{\Lambda} \cdot \mathbf{R} = \mathbf{P}^{\mathrm{T}} \cdot \tilde{\mathbf{F}}$$

with initial conditions

$$\mathbf{R}(0) = \begin{bmatrix} \mathbf{P}^T \cdot \mathbf{M}^{1/2} \cdot x_{10} \\ \mathbf{P}^T \cdot \mathbf{M}^{1/2} \cdot x_{20} \end{bmatrix}$$

After denoting

$$\mathbf{\Lambda} = \mathbf{\Omega}^2 = \begin{bmatrix} \Omega_1^{\;2} & 0 \\ 0 & \Omega_2^{\;2} \end{bmatrix}$$

and given

$$\mathbf{P}^T = \begin{bmatrix} \mathbf{V}_1^T \\ \mathbf{V}_2^T \end{bmatrix} = \begin{bmatrix} v_{11} & v_{12} \\ v_{21} & v_{22} \end{bmatrix}$$

we obtain

$$\mathbf{P}^T \tilde{\mathbf{F}} = \begin{bmatrix} \mathbf{V}_1^{\;T} \\ \mathbf{V}_2^{\;T} \end{bmatrix} \begin{bmatrix} m_1^{\;-1/2} \cdot f_1 \\ 0 \end{bmatrix} = \begin{bmatrix} v_{11} & v_{12} \\ v_{21} & v_{22} \end{bmatrix} \begin{bmatrix} m_1^{\;-1/2} \cdot f_1 \\ 0 \end{bmatrix} = \begin{bmatrix} v_{11} \cdot m_1^{\;-1/2} \cdot f_1 \\ v_{21} \cdot m_1^{\;-1/2} \cdot f_1 \end{bmatrix}$$

The above matrix equation of motion can be written in scalar form as

$$\ddot{r}_1(t) + \Omega_1^{\;2} \cdot r_1(t) = m_1^{\;-1/2} \cdot f_1(t) \cdot v_{12}$$

$$\ddot{r}_2(t) + \Omega_2^{\;2} \cdot r_2(t) = m_1^{\;-1/2} \cdot f_1(t) \cdot v_{21}$$

These are decoupled modal equations of the above under-actuated 2 DOF mechanical system. Figure 5.13 shows the decoupled modes, equivalent to the system from Fig. 5.12.

It can be observed that, even if only the mass m_1 in Fig. 5.12 is subject to the external undefined force $f_1(t)$ and m_2 is subject to no force, both unit masses for modes 1 and 2 are actuated, but cannot be independently controlled by the single force $f_1(t)$. Choosing feedback control to determine $f_1(t)$ for mode 1, for mode 2, $v_{21} \cdot m_1^{-1/2} \cdot f_1$ cannot be modulated any more to satisfy control needs for mode 2. The term $v_{21} m_1^{-1/2} f_1$ will however excite mode 2 as a spillover.

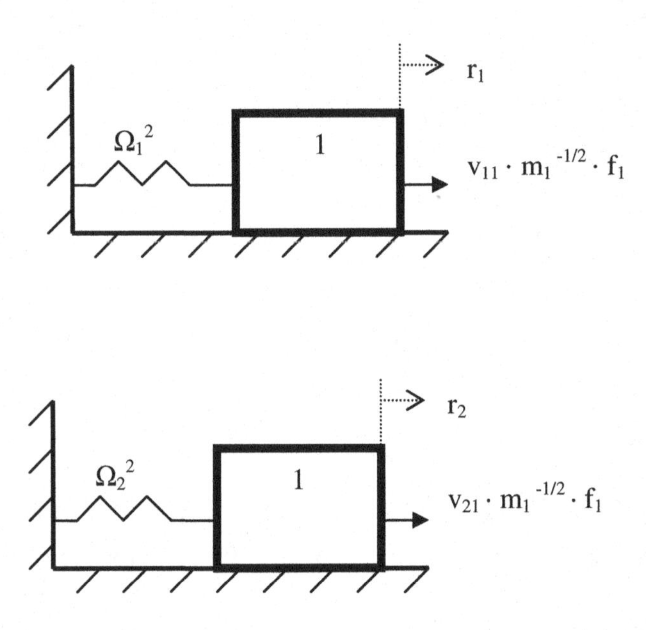

Fig. 5.13 Decoupled modal equivalent of the mechanical system from Fig. 5.12

5.3.5 *Modal Control in Beam Vibration*

The generic solution to the direct problem, presented in the previous section as direct problem for beam vibration modeling as a double infinite series cannot be used for solving analytically the inverse problem that would give the excitation f(x, t) that would produce the desired transversal displacement $y_d(x, t)$. Consequently, this non-closed form solution cannot be used for obtaining a feed-forward control law. This solution was however used for the formulation of the feedback control in the form of reduced order modal control, presented in the specialized literature [17, 23].

The first three modes, they contain $\sin x \cdot (\pi / l)$, $\sin 2 \cdot (\pi / l)$ and $\sin 3x \cdot (\pi / l)$ give the shapes shown in Fig. 5.14. The number of nodes for mode n is n - 1. Modal control can be implemented, for maximum efficiency, with actuators located mid-distance between adjacent nodes, i.e. $x = l / 2$ for $n = 1$, $x = l / 4$ or $3 \cdot l / 4$ for $n = 2$ and $x = l / 6$, $l / 2$ or $5 \cdot l / 6$ for $n = 3$. This shows that there are required as many different actuators as the number of modes desired to control. Space, cost and design constraints limit the number of modes that can be controlled to, normally, less than 10. Modes that are not controlled but can be excited by the actuators outputs for the modes intended to control result in spillover phenomenon that has to be addressed separately [23].

5.4 Direct Problem in Free Vibrations in Membranes

The membrane shown in Fig 5.15 has a small transversal (along z axis) displacement u(x, y, t), in the plane x, y from the equilibrium position [17, 25, 109].

The equation for free vibrating membrane for transversal displacement u(x, y, t) is [17, 23, 35]

$$(\overline{m}/(\sigma \cdot h))\frac{\partial^2 u(x,y,t)}{\partial t^2} = \frac{\partial^2 u(x,y,t)}{\partial x^2} + \frac{\partial^2 u(x,y,t)}{\partial y^2}$$

where

$\overline{m}$ is mass per unit area
$\sigma \cdot h$ is the uniform tensile force

The solution can be obtained using the method of separation of variables [14, 17]

$$u(x, y, t) = W(x, y) \cdot \theta(t)$$

where W(x, y) is mode shape function and θ(t) is the free vibration time dependence.

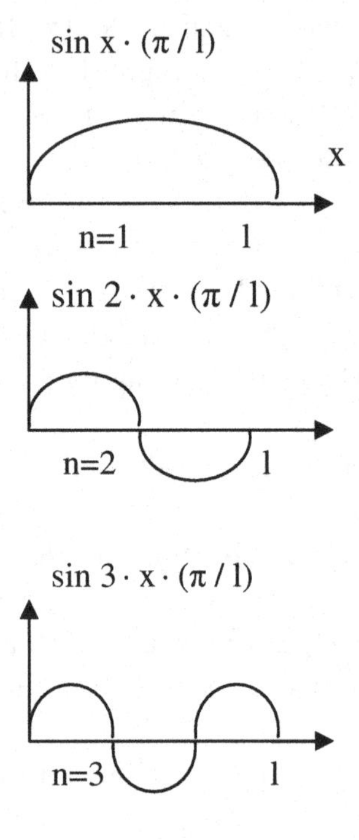

Fig. 5.14 The first three mode shapes for a simply supported beam

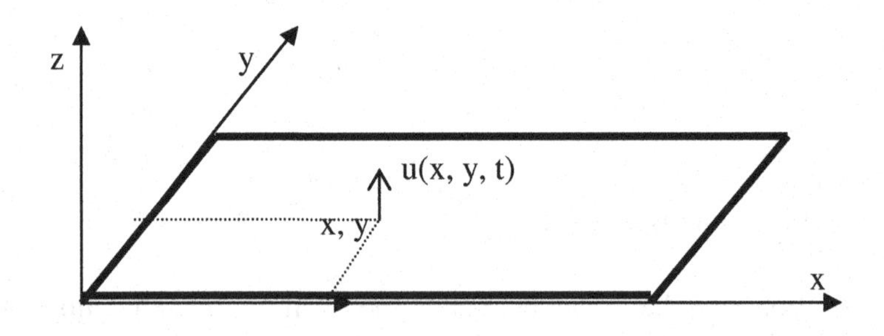

Fig. 5.15 Membrane vibrating transversally

Assuming a harmonic function $\theta(t)$ of unit amplitude frequency ω

$$\theta(t) = \sin \omega t$$

and a double sine series representation for W(x, y) for a simply supported rectangular membrane a by b with fixed boundaries

$$W(x, y) = \sum_{m=1}^{\infty} \sum_{n=1}^{\infty} W_{m,n} \cdot \sin(n \cdot \pi \cdot x / a) \cdot \sin(n \cdot \pi \cdot x / b)$$

membrane equation becomes

$$\pi^2 \cdot m^2 / a^2 + \pi^2 \cdot n^2 / b^2 = \omega^2 (\overline{m} / (\sigma \cdot h))$$

with the solution for the radial frequency of the mode $m \cdot n$

$$\omega_{mn} = \pi \sqrt{[(m^2 / a^2 + n^2 / b^2) / (\overline{m} / (\sigma \cdot h))]}$$

Similar to the case of the beam, the direct problem solution for the membranes cannot be used for obtaining the feed-forward control law. This direct problem solution, a triple infinite series, can be plotted using MAPLE [25] and can be simulated using FEMLAB [36].

5.4.1 *Membrane Vibration Solution Plotting*

Membrane vibration solution plotting using MAPLE shown in Fig. 5.16 a to e is obtained for:
-initial conditions:

$$u(x, y, 0) = \Phi(x, y)$$
$$du(x, y, 0)/dt = \Psi(x, y)$$

-boundary conditions for the membrane fixed at four corners:
for $t > 0$, $u = 0$ for the four corners and for $t = 0$, $du / dt = 0$.

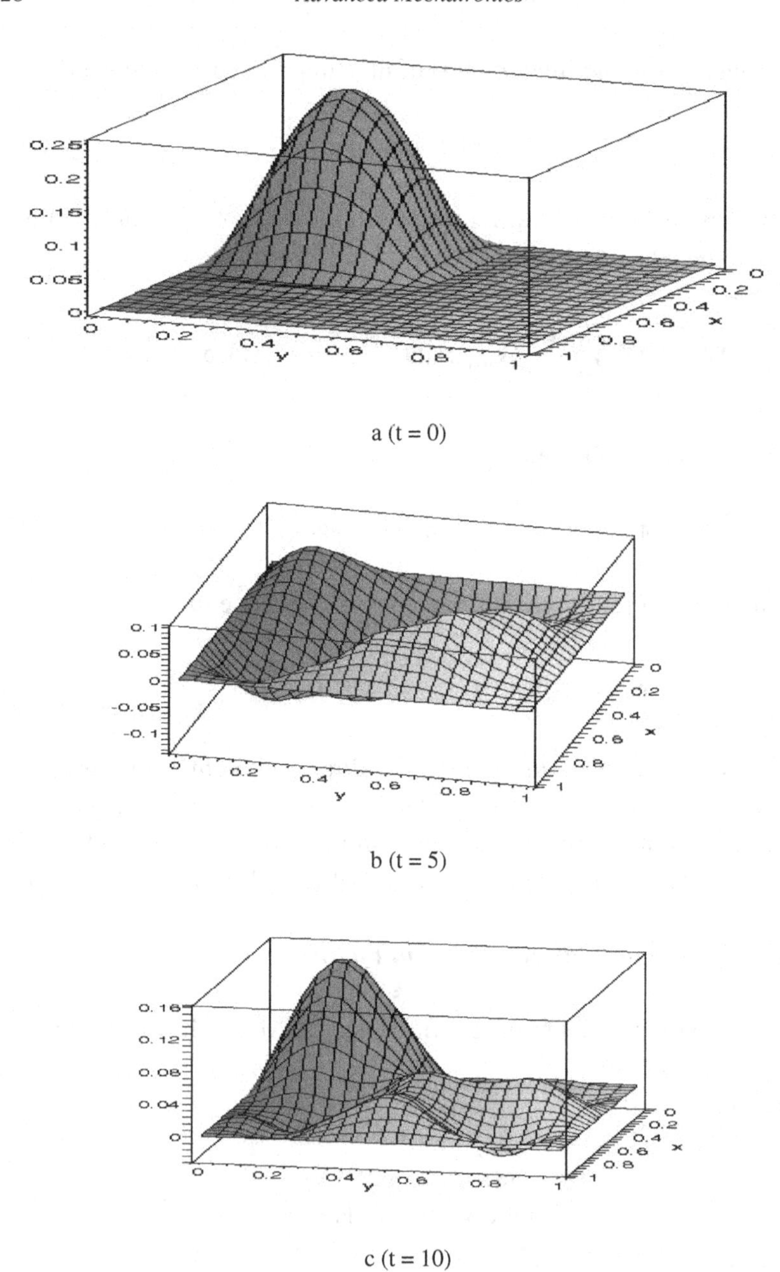

a (t = 0)

b (t = 5)

c (t = 10)

Fig. 5.16 MAPLE simulation results for membrane transversal vibrations

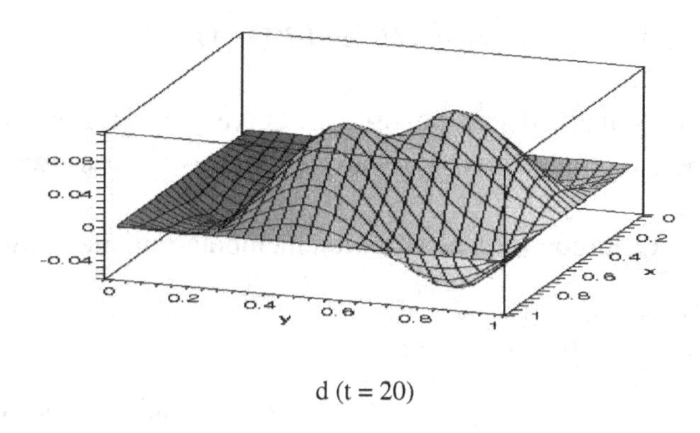

d (t = 20)

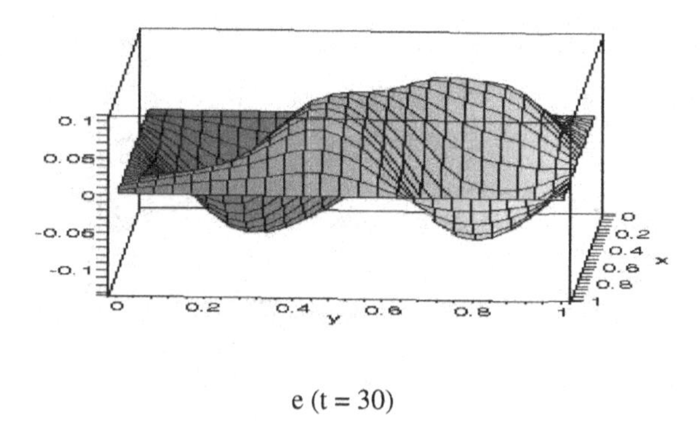

e (t = 30)

Fig. 5.16 (*Continued*)

These results for the direct problem of free membrane vibration, subject to non-zero initial conditions, show even in qualitative analysis the complexity of u(x, y, t) shape change over time. Inverse problem consists in obtaining the required input forces to achieve a desired output $u_d(x, y, t)$. Real-time application of an inverse problem solution is not practical at this time.

5.4.2 *Simulation of Membrane Using FEMLAB*

Finite elements method can be used to solve numerically membrane equation [36]. A commercial software, FEMLAB™, was used for this purpose.

Boundary condition for free vibration membrane simulation were:

-Fixed at left and right.
-Front and back are free to vibrate.

The results are shown in Fig. 5.17 a to e. Initial conditions are nonzero, as shown in Fig. 5.17 a.

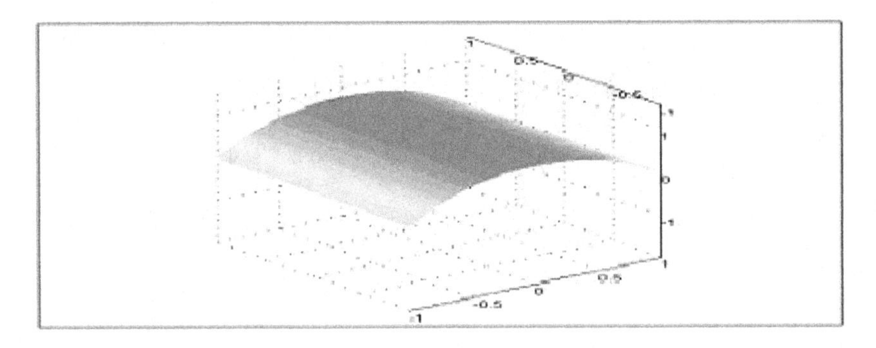

a (t = 0)

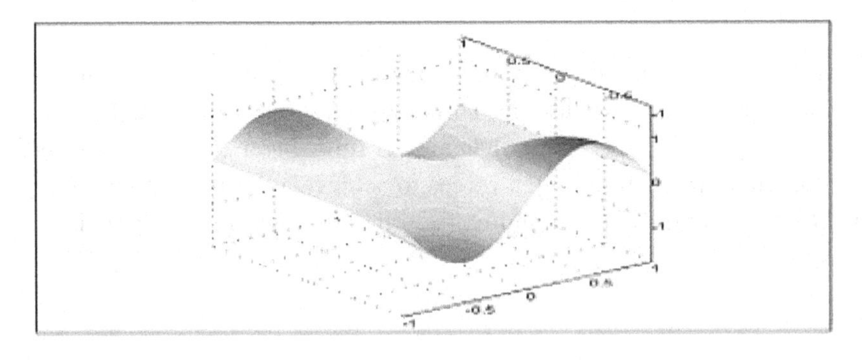

b (t = 5)

Fig. 5.17 FEMLAB simulation results for membrane transversal vibrations

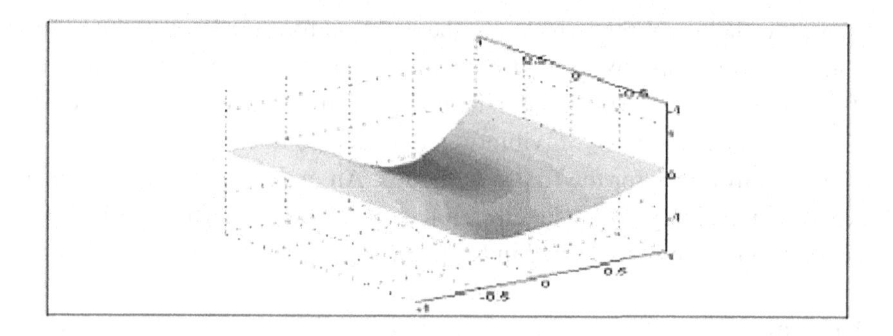

c (t = 10)

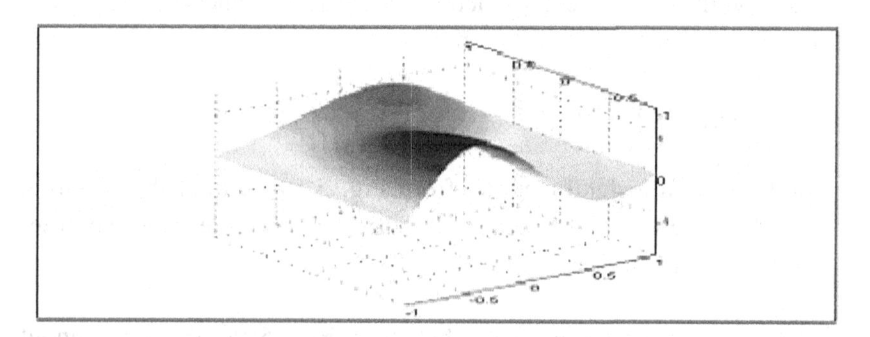

d (t = 20)

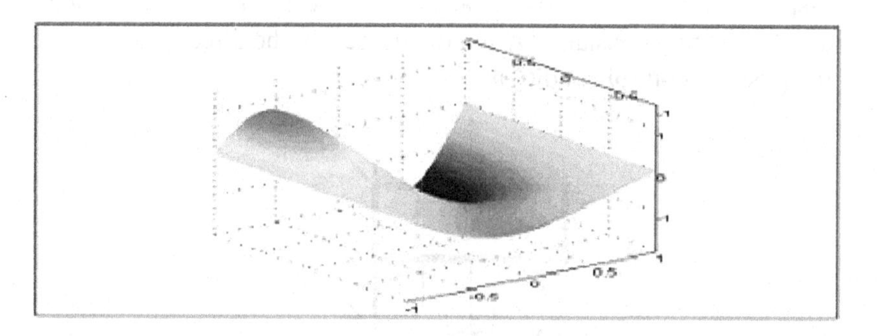

e (t = 30)

Fig. 5.17 (*Continued*)

Similar to the membrane vibration, the solution for the direct problem for plate vibration consists in multiple sums of infinite series for an infinity of vibration modes and results in complex time varying shapes [55]. Also, feedback control of vibration displacement can be achieved in modal control only for the first few modes. An example of analytical and experimental study of plate vibration control reduced to the first two modes of vibration is reported in [23].

Problems

1. Assume the experimentally determined transfer functions, limited to the first two modes of vibration

$$O_m(\omega) / I^{(c)}(\omega) = k \cdot (s - z) \cdot e^{-T\omega} / [(s + p_1) \cdot (s + p_2)]$$

for the tree transfer functions from Fig. 5.10 Obtain the resulting transfer function of the feedback controller and determine the order of its denominator.

2. For the mechanical system, shown in Fig. 5.18, the force input subject to control is f(t), the external force perturbation is f_{ext} and the displacement output is y(t). B and b are viscous friction coefficients and K is spring constant. Obtain the model for the direct problem and the position control condition.

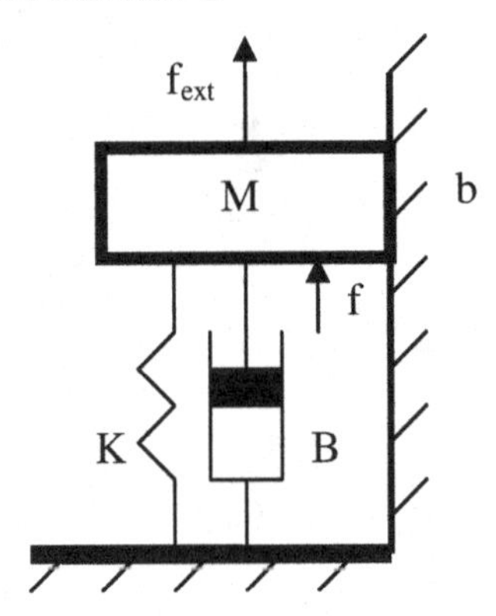

Fig. 5.18 Mechanical system

3. For the mechanical system, shown in Fig. 5.19, obtain the transfer function for the direct problem and the block diagram for PID position control.

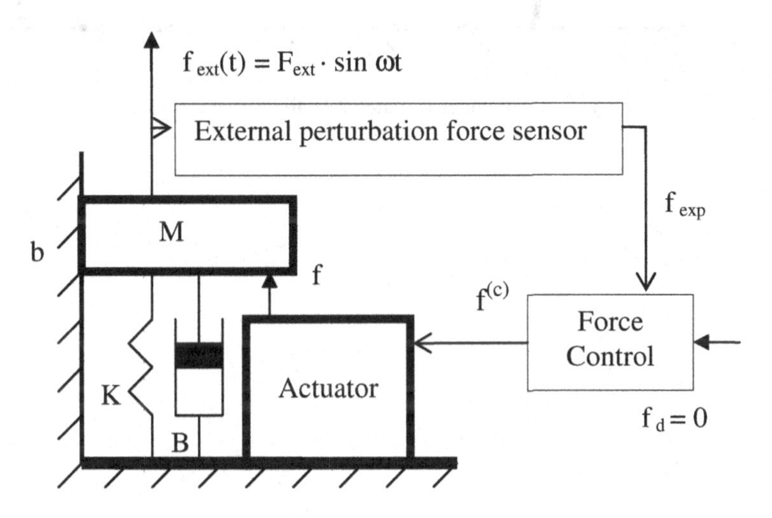

Fig. 5.19 Position control diagram for vibration suppression

4. Assume a flexible shaft represented by a single finite element model. For the case that $T_1 = 0$ and $T_2 = \tau_2$, draw the block diagram for the closed loop control to bring $\theta_2(s)$ towards $\theta_2(s)$. Show the relationship between $\theta_1(s)$ and $\theta_2(s)$.

5. For the under-actuated 2 DOF mechanical system, shown in Fig. 5.20, obtain the decoupled modal equivalent

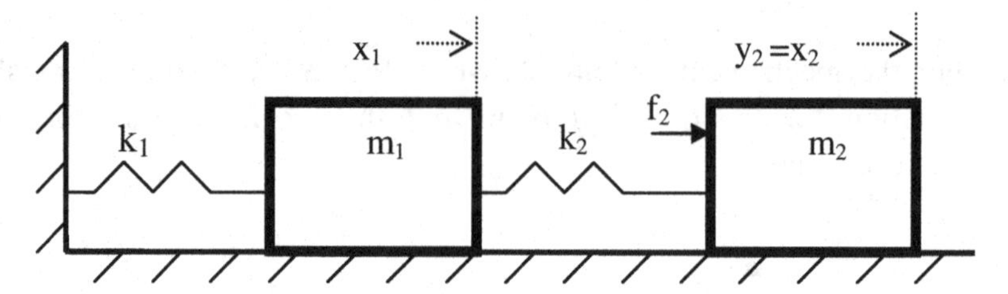

Fig. 5.20 An under-actuated 2 DOF mechanical system

Chapter 6

Acousto-Mechatronics

6.1 Acousto-Mechatronic Systems

Distributed parameters acousto-mechatronic systems are mixed systems containing acoustic field transmission under computerized tight integration. Such systems can function only under permanent computer monitoring and control of the state variables of the acoustic field [82-84]. Similar to other Distributed Parameters Systems (DPS), a finite number of point sensors and actuators result in an under-sensed and under-actuated acoustic system. This chapter focuses on room acoustics and the use of direct and reflected ray propagation in discrete inverse problems solving for parameters estimation.

6.1.1 *Recording Studio*

Figure 6.1 shows the conceptual diagram of a sound recording studio. Acoustic signals from voice, musical instruments and other sound sources to be recorded are assumed transmitted through the enclosed space of a room (recording studio), in fact a Distributed Parameters System. These acoustic signals, modified by the room acoustics (reverberations, wall sound absorption etc) are inputs to microphone(s) or microphones, i.e. sensors converting acoustic signals into modulated voltage signals [21, 85]. A recording system records these signals on hard-drives, tapes, CDs etc. Often recoding takes place in an un-echoic studio in order to avoid reflected waves. Sound effects in records can be, however, added by digital signal post-processing.

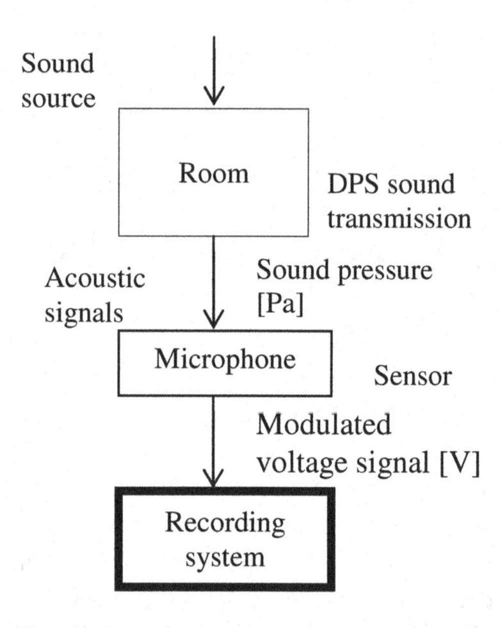

Fig. 6.1 Conceptual diagram of a recording studio

6.1.2 *Active Sound Control in Halls*

Figure 6.2 shows the conceptual diagram of a room (hall) active acoustics. Modulated voltage signals from, for example, microphone on a stage or from sound recording, provide inputs to speakers (i.e. actuators) that generate acoustic signal outputs, transmitted further in a room (hall), assumed an enclosed space modeled as a DPS. These acoustic signals, modified by the room acoustics (reverberations, wall sound absorption etc) arrive to audience ears, i.e. to acoustic receivers [21, 85-87].

This system will also be analyzed in this section as a distributed parameters acousto-mechatronic system.

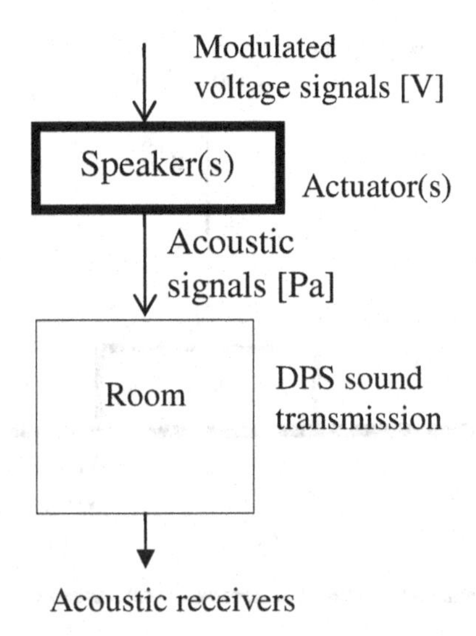

Fig. 6.2 Conceptual diagram of a room (hall) active acoustics

6.1.3 *Active Noise Control*

Figure 6.3 shows the active control scheme for noise control in a duct. The external perturbation is, in this case, a noise source of pressure $w(x_e, t)$ at x_e, that propagates in a duct. The noise is measured with a microphone placed at x_e, that has the output $w_m(x_e, t)$. A feedback controller produces the analog voltage command $u^{(c)}$ sent to a speaker. The sound generated by the speaker, $y(x_u, t)$, combines with the noise and both propagate in the duct. A monitoring microphone, placed at x_e, produces and output voltage $E(x_d, t)$ which, in the case of ideal active (feed-forward) control, is supposed to be zero [23].

This negative feedback system, with non-collocated sensor, actuator and output is designed to make the signal $E(x_d, t)$ tend toward its desired zero value.

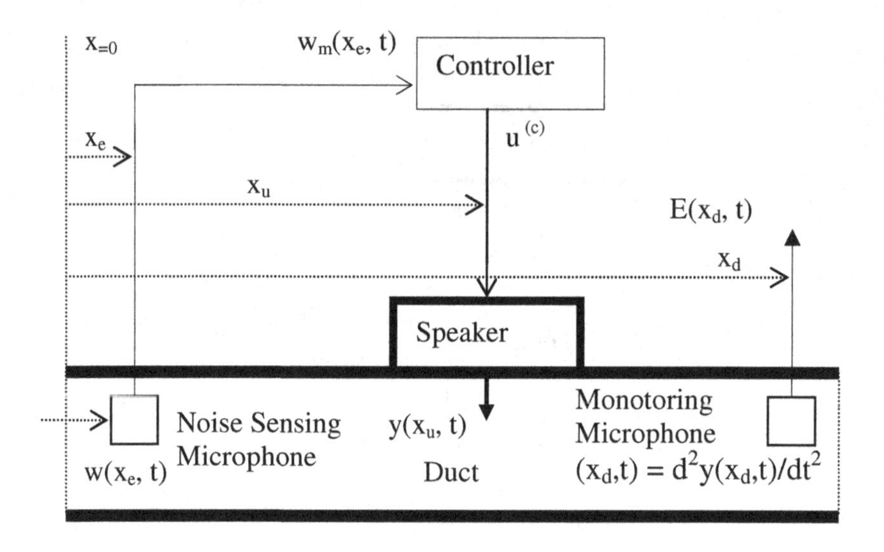

Fig. 6.3 Active noise control in a duct

6.2 Distributed Parameters Models of Sound Transmission

6.2.1 Wave Equation for Planar Sound Wave 1D Propagation in a Free Sound Field

A simple case of acoustic wave propagation of the plane sound wave propagating in a free (nonreflecting) 3D space is shown in Fig. 6.4. Assuming a wave propagating in x direction from a source located x_s at constant sound speed c, the equation for the plane sound wave (i.e. far from the source, such that spherical waves can be approximated by planar waves) is the same as the equation for vibrating string [17, 21, 39]

$$\frac{\partial^2 u(x,t)}{\partial t^2} = c^2 \frac{\partial^2 u(x,t)}{\partial x^2}$$

where

$u(x, t)$ is the displacement of the longitudinal sound wave in the positive and negative direction about an equilibrium position

$$u_e(x, t) = 0$$

and

$$c = \sqrt{\beta} / \rho_0$$

β is the bulk modulus (modulus of elasticity of the medium)
ρ_0 is the density

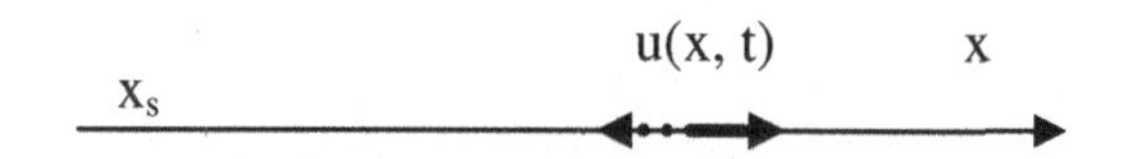

Fig. 6.4 Planar sound wave propagation

Assuming that the source, seen from a far point x, can be represented as a planar wave source generating a simple harmonic motion with frequency

$$\omega = 2 \cdot \pi \cdot f = 2 \cdot \pi / T$$

then

$$u = U_s \cdot \cos \omega \cdot t$$

and the plane traveling waves with velocity c arrive at a location x after x / c, such that the displacement $u(x, t)$, in case of no attenuation, is given by

$$u(x, t) = U_s \cdot \cos \omega \cdot (t - x / c)$$

or

$$u(x, t) = U_s \cdot \cos(\omega \cdot t - k \cdot x)$$

where the wave-number k is

$$k = \omega / c = 2 \cdot \pi / \lambda$$

and the wavelength is

$$\lambda = 2 \cdot \pi \cdot c / \omega$$

The proposed solution is

$$u(x, t) = U_s \cdot \cos \omega \cdot (t - x / c)$$

with

$$\frac{\partial^2 u(x,t)}{\partial t^2} = -U_s \cdot \omega^2 \cdot \cos(\omega \cdot t - k \cdot x)$$

while

$$\frac{\partial^2 u(x,t)}{\partial x^2} = -U_s \cdot k^2 \cdot \cos(\omega \cdot t - k \cdot x)$$

is a solution of the sound wave displacement equation

$$\frac{\partial^2 u(x,t)}{\partial t^2} = c^2 \frac{\partial^2 u(x,t)}{\partial x^2}$$

where, as above,

$$c = k / \omega$$

The plane wave of the longitudinal vibration of the air about an equilibrium position can be interpreted as a traveling sound pressure wave, which is a variation of the pressure, associated with sound propagation, about the local static value of the atmospheric pressure generally (approx. 10,000 Pa). For the free (nonreflecting) sound propagation case, this wave propagates in the positive x-direction, such that any value U of the wave $U = U_s \cdot \cos(\omega \cdot t - k \cdot x)$ can be seen as traveling rightwards with velocity c.

Stress-strain Hook equation, (in this case for pressure p and strain $\delta u(x, t) / \delta x$) can be written as

$$p = -\beta \cdot \delta u(x, t) / \delta x$$

showing that a layer of the propagation medium is compressed by a positive p and the strain which explains the negative sign

$$\delta u(x, t) / \delta x = -p / \beta$$

For the above solution of the sound wave displacement equation

$$u(x, t) = U_s \cdot \cos(\omega \cdot t - k \cdot x)$$

the time derivative is

$$\delta u(x, t)/\delta x = U_s \cdot k \cdot \sin(\omega \cdot t - k \cdot x)$$

and the pressure solution becomes

$$p = -\beta \cdot \delta u(x, t) / \delta x = -\beta \cdot k \cdot U_s \cdot \sin(\omega \cdot t - k \cdot x)$$

such that, for

$$c = \pm\sqrt{\beta} / \rho_0$$

or

$$\beta = c^2 \cdot \rho_0$$

and

$$k = \omega / c$$

gives

$$p = -\beta \cdot k \cdot U_s \cdot \sin(\omega \cdot t - k \cdot x) = -c \cdot \omega \cdot \rho_0 \cdot U_s \cdot \sin(\omega \cdot t - k \cdot x)$$

Second derivative, with regard to time, is

$$\frac{\partial^2 p(x,t)}{\partial t^2} = (c \cdot \omega \cdot \rho_0 \cdot U_s) \cdot \omega^2 \cdot \sin(\omega \cdot t - k \cdot x)$$

while, with regard to x, is

$$\frac{\partial^2 u(x,t)}{\partial x^2} = (c \cdot \omega \cdot \rho_0 \cdot U_s) \cdot k^2 \cdot \sin(\omega \cdot t - k \cdot x)$$

Sound pressure wave equation gives [21, 57]

$$\frac{\partial^2 p(x,t)}{\partial t^2} = c^2 \frac{\partial^2 p(x,t)}{\partial x^2}$$

or

$$\frac{\partial^2 p(x,t)}{\partial t^2} - c^2 \frac{\partial^2 p(x,t)}{\partial x^2} = 0$$

For, $c = \pm\sqrt{\beta} / \rho_0$, both $U_s \cdot \sin(\omega \cdot t - k \cdot x)$ and $U_s \cdot \sin(\omega \cdot t + k \cdot x)$ verify the wave equation.

For an enclosed propagation medium, like a finite length duct, incident waves propagating in the + x direction $U_s \cdot \cos(\omega \cdot t - k \cdot x)$ interfere with the reflecting waves propagating in the – x direction $U_s \cdot \cos(\omega \cdot t + k \cdot x)$, (ignoring for now phase difference), forming standing waves

$$U_s \cdot \cos(\omega \cdot t - k \cdot x) + U_s \cdot \cos(\omega \cdot t + k \cdot x) = 2 \cdot U_s \cdot (\cos k \cdot x) \cdot (\cos \omega \cdot t)$$

i.e. non propagating waves (given that there is no factor $\omega \cdot t \pm k \cdot x$), oscillating locally at x, with amplitude $2 \cdot U_s \cdot (\cos k \cdot x)$ and with the time variation $\cos \omega t$ [60].

6.2.2 *Wave Equation for Planar Sound Wave 3D Propagation a Free Sound Field*

In this section is analyzed the case of acoustic wave propagation as a plane sound wave propagating in a free (nonreflecting) 3D space, shown in Fig. 6.5.

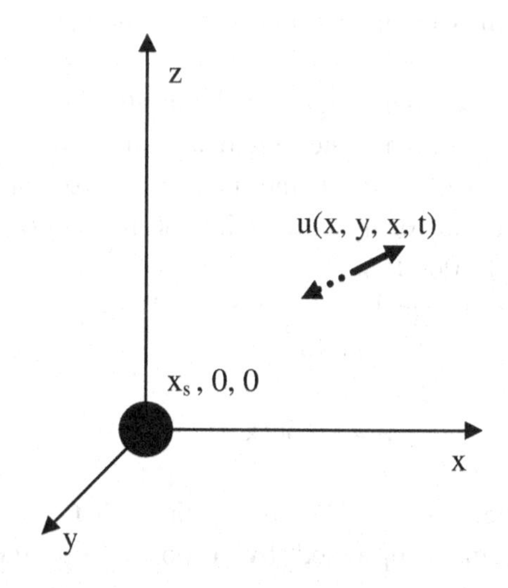

Fig. 6.5 Planar sound wave propagation in 3D free space

The wave propagates in (x, y, z) space from a source located $(x_s, 0, 0)$ at constant sound speed c. The equation for the plane sound wave displacement u(x, y, z, t) is

$$\frac{\partial^2 u(x,y,z,t)}{\partial t^2} - c^2 (\frac{\partial^2 u(x,y,z,t)}{\partial x^2} + \frac{\partial u^2 (x,y,z,t)}{\partial y^2} + \frac{\partial u^2 (x,y,z,t)}{\partial z^2}) = 0$$

or, in compact notation,

$$\frac{\partial^2 u(x,y,z,t)}{\partial t^2} - c^2 \cdot \Delta u(x,y,z) = 0$$

where

$u_e(x, y, z, t) = 0$ is the displacement of the longitudinal sound wave in the positive and negative direction about a equilibrium position, and

$$c = \sqrt{\beta} / \rho_0$$

β is the bulk modulus (modulus of elasticity of the medium)
ρ_0 is air density.

The plane wave of the longitudinal vibration of the air about an equilibrium position can be interpreted as a traveling sound pressure wave, which is a variation of the pressure, associated with sound propagation, about the local static value of the atmospheric pressure generally (approx. 10,000 Pa).

Stress-strain Hook equation, (in this case for pressure p and strain $\delta u(x, t) / \delta x$) is

$$p = - \beta\ \delta u(x, t) / \delta x$$

showing, as in the case of 1D propagation, that for a layer of the propagation medium compressed by a positive p, the strain has a negative value.

Based on this, sound pressure p (x, y, z, t) wave equation results as [21, 82, 83]

$$\frac{\partial^2 p(x,y,z,t)}{\partial t^2} - c^2 (\frac{\partial^2 p(x,y,z,t)}{\partial x^2} + \frac{\partial p^2 (x,y,z,t)}{\partial y^2} + \frac{\partial p^2 (x,y,z,t)}{\partial z^2})$$

$$= 0$$

or, in compact notation

$$\frac{\partial^2 p(x,y,z,t)}{\partial t^2} - c^2 \cdot \Delta p(x,y,z,t) = 0$$

Assuming a planar wave source generating a simple harmonic motion with frequency $\omega = 2 \cdot \pi \cdot f = 2 \cdot \pi/T$

$$u = U_s \cdot \cos \omega t$$

a plane traveling waves along x-coordinate with velocity c arrive at a location x after x/c, and the displacement u(x, t), in case of no attenuation, is given by

$$u(x, t) = U_s \cdot \cos \omega(t - x / c) = U_s \cdot \cos (\omega \cdot t - k \cdot x)$$

where the wave-number k is

$$k = \omega / c = 2 \cdot \pi / \lambda$$

and λ is the wavelength

6.2.3 *Sound Wave Propagation in an Enclosed Sound Field*

The above wave equations can be used for modeling the sound propagation in an enclosed space, by defining the specific boundary conditions.

For example [56]

-for hard boundaries, (walls), with surface normal vector n

$$\frac{\partial p}{\partial n} = 0$$

-for a pressure source of constant pressure p_0, located on the boundary

$$p = p_0$$

Numerical solvers for partial differential equations, for example FEMLAB™, give the numerical solutions for the sound wave equation for the particular boundary conditions for the enclosed space [56, 57].

6.3 Calculation of Room Eigenvalues and Eigenvectors for a Rectangular Cavity or Room

A rectangular cavity of dimensions X, Y and Z is shown in Fig. 6.6.

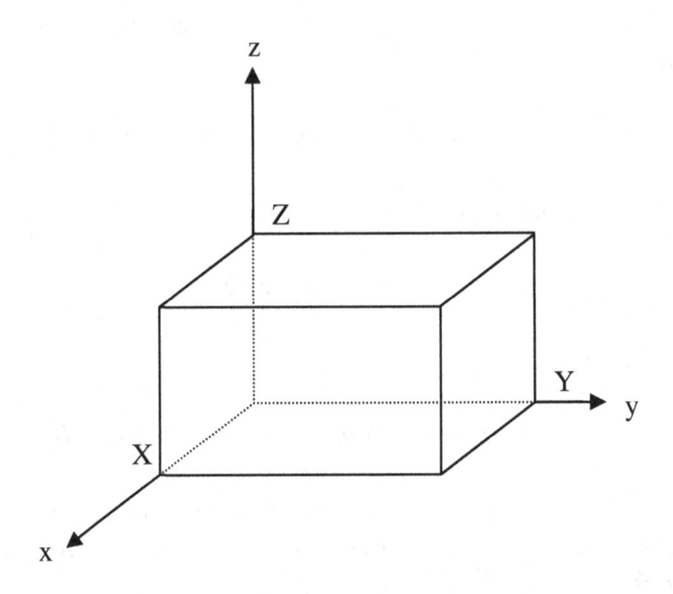

Fig. 6.6 Rectangular enclosed space XYZ

The calculation of eigenvalues is based on the sound wave equation for an assumed time harmonic solution [57]

$$p(x, y, z, t) = P(x, y, z) \cdot e^{j\omega t}$$

such that

$$\frac{\partial^2 p(x,y,z,t)}{\partial t^2} = -\omega^2 \cdot P(x,y,z)e^{j\omega t}$$

and

$$\Delta p(x,y,z) = e^{j\omega t} \cdot \Delta P(x,y,z)$$

The pressure wave equation gives

$$-\omega^2 \cdot P(x,y,z)e^{j\omega t} - c^2 \cdot e^{j\omega t} \cdot \Delta P(x,y,z) = 0$$

This results in the Helmholz equation [21, 57]

$$\Delta P(x,y,z) + (\omega / c)^2 \cdot P(x,y,z) = 0$$

or

$$\frac{\partial^2 P(x,y,z)}{\partial x^2} + \frac{\partial^2 P(x,y,z)}{\partial y^2} + \frac{\partial^2 P(x,y,z)}{\partial z^2} + k^2 \cdot P(x,y,z) = 0$$

where $k = \omega / c$.

The method of separation of variables for P(x, y, z)

$$P(x, y, z) = A(x) \cdot B(y) \cdot C(z)$$

gives

$$\frac{\partial^2 A(x)}{\partial x^2} B(y) \cdot C(z) + \frac{\partial^2 B(y)}{\partial y^2} A(x) \cdot C(Z) + \frac{\partial^2 C(z)}{\partial z^2} A(x) \cdot B(y)$$
$$+ k^2 \cdot A(x) \cdot B(y) \cdot C(z) = 0$$

where, for 3D propagation, the wave number 3D vector **k** is

$$\mathbf{k} = k_x \cdot i + k_y \cdot j + k_z \cdot k$$

and $|\mathbf{k}|^2 = k^2$

$$k^2 = k_x^2 + k_y^2 + k_z^2 = (\omega / c)^2$$

or

$$\omega = c \sqrt{(k_x^2 + k_y^2 + k_z^2)}$$

After dividing by $A(x) \cdot B(y) \cdot C(z)$, the following equation results

$$\frac{d^2A(x)}{A(x) \cdot dx^2} + \frac{d^2B(y)}{B(y) \cdot dy^2} + \frac{d^2C(z)}{C(z) \cdot dz^2} + k_x^2 + k_y^2 + k_z^2 = 0$$

or

$$(\frac{d^2A(x)}{A(x) \cdot dx^2} + k_x^2) + (\frac{d^2B(y)}{B(y) \cdot dy^2} + k_y^2) + (\frac{d^2C(z)}{C(z) \cdot dz^2} + k_z^2) = 0$$

The three terms in this equation are equal to zero for all x, y and z only if each separate term is equal to zero

$$(\frac{d^2A(x)}{A(x) \cdot dx^2} + k_x^2) = 0$$

$$(\frac{d^2B(y)}{B(y) \cdot dy^2} + k_y^2) = 0$$

$$(\frac{d^2C(z)}{C(z) \cdot dz^2} + k_z^2) = 0$$

For the nontrivial case that $A(x)$, $B(y)$ and $C(z)$ are all non-zero, three separate second order ordinary differential equations result [21]

$$(\frac{d^2}{dx^2}+k_x{}^2)\cdot A(x)=0$$

$$(\frac{d^2}{dy^2}+k_y{}^2)\cdot B(y)=0$$

$$(\frac{d^2}{dz^2}+k_z{}^2)\cdot C(z)=0$$

These equations suggest harmonic spatial solutions, for example sin or cos types. For the cos type, the solutions could be cos $k_{xl}\cdot x$, cos $k_{ym}\cdot y$ and cos $k_{zn}\cdot z$, for l. m, n = 0,1,2,3,…

Consider a room with all boundaries, *i.e.* all walls, ceiling and floor, assumed as hard boundaries,

$$\frac{\partial p(x,y,z,t)}{\partial x}=0 \quad \text{for} \quad x=0 \quad \text{and} \quad x=X$$

$$\frac{\partial p(x,y,z,t)}{\partial y}=0 \quad \text{for} \quad y=0 \quad \text{and} \quad y=Y$$

$$\frac{\partial p(x,y,z,t)}{\partial z}=0 \quad \text{for} \quad z=0 \quad \text{and} \quad z=Z$$

For the wave equation

$$\frac{\partial^2 p(x,y,z,t)}{\partial t^2}-c^2\cdot\Delta p(x,y,z,t)=0$$

and for an assumed time harmonic solution [57]

$$p(x, y, z, t) = P(x, y, z)\cdot e^{j\omega t}$$

the spatial part P(x, y, z) has to be formed of cos terms, such that lmn solutions for p(x, y, z,) could be

$$p_{lmn}(x, y, z, t) = A_{lmn} \cdot (\cos k_{xl} \cdot x) \cdot (\cos k_{ym} \cdot y) \cdot (\cos k_{zn} \cdot z) \cdot e^{j\omega t}$$

for l, m, n = 0,1,2,3,... that can be made to verify the above hard boundary conditions *i.e.*

$$\delta p(x, y, z, t) / \delta x = -A_{lmn} \cdot k_{xl} \cdot (\sin k_{xl} \cdot x) \cdot (\cos k_{ym} \cdot y) \cdot (\cos k_{zn} \cdot z) \cdot e^{j\omega t} = 0 \quad \text{for } x = 0 \text{ and } x = X$$

$$\delta p(x, y, z, t) / \delta y = -A_{lmn} \cdot k_{yl} \cdot (\cos k_{xl} \cdot x) \cdot (\sin k_{ym} \cdot y) \cdot (\cos k_{zn} \cdot z) \cdot e^{j\omega t} = 0 \quad \text{for } y = 0 \text{ and } y = Y$$

$$\delta p(x, y, z, t) / \delta y = -A_{lmn} \cdot k_{zl} \cdot (\cos k_{xl} \cdot x) \cdot (\sin k_{ym} \cdot y) \cdot (\sin k_{zn} \cdot z) \cdot e^{j\omega t} = 0 \quad \text{for } z = 0 \text{ and } z = Z$$

In case of $x = 0$, $y = 0$ and $z = 0$, the above conditions are satisfied due to the factors $\sin (k_{xl} \cdot 0) = 0$, $\sin (k_{ym} \cdot 0) = 0$ and $\sin (k_{zn} \cdot 0) = 0$, respectively.

In case of $x = X$,

$$\delta p(x, y, z, t) / \delta x = -A_{lmn} \cdot k_{xl} (\sin k_{xl} \cdot X) \cdot (\cos k_{ym} \cdot y) \cdot (\cos k_{zn} \cdot z) \cdot e^{j\omega t} = 0$$

is satisfied when $\sin (k_{xl} \cdot X) = 0$, or when

$$k_{xl} = \pi \cdot l / X \quad \text{for } l = 0,1,2,3,\ldots$$

Similarly, for the other two boundary conditions

$$k_{ym} = \pi \cdot m / Y \quad \text{for } m = 0,1,2,3,\ldots$$
$$k_{zn} = \pi \cdot n / Z \quad \text{for } n = 0,1,2,3,\ldots$$

This gives

$$k_{lmn} = \sqrt{(k_x^2 + k_y^2 + k_z^2)} = \pi \sqrt{[(l/X)^2 + (m/Y)^2 + (n/Z)^2]}$$
$$\text{for l. m, n} = 0,1,2,3,\ldots$$

and

$$\omega_{lmn} = c \cdot \pi \cdot \sqrt{[(l/X)^2 + (m/Y)^2 + (n/Z)^2]} = 2 \cdot \pi \cdot f_{lmn}$$
$$\text{for l. m, n} = 0,1,2,3,\ldots$$

or, modal frequencies

$$f_{lmn} = (c/2) \cdot \sqrt{[(l/X)^2 + (m/Y)^2 + (n/Z)^2]}$$
$$\text{for l. m, n} = 0,1,2,3,\ldots$$

This shows that larger room dimensions, X, Y, Z lead to lower modal frequencies f_{lmn}.

Example 6.1 For a rectangular room, as shown in Fig. 6.6, with dimensions X = 2 [m], Y = 1 [m] and Z = 1[m], calculate natural frequencies for the first few modes, in case of sound speed in air c = 343.6 [m/s].

Natural frequencies are given by

$$f_{lmn} = (171.8) \cdot \sqrt{[(l/2)^2 + (m)^2 + (n)^2]}$$
$$\text{for l. m, n} = 0,1,2,3,\ldots$$

such that

$f_{000} = (171.8) \sqrt{[(0/2)^2 + (0)^2 + (0)^2]} = 0$ i.e. no sound transmission
$f_{100} = (171.8) \sqrt{[(1/2)^2 + (0)^2 + (0)^2]} = 85.9$ [Hz]
$f_{010} = (171.8) \sqrt{[(0/2)^2 + (1)^2 + (0)^2]} = 171.8$ [Hz]
$f_{001} = (171.8) \sqrt{[(0/2)^2 + (0)^2 + (1)^2]} = 171.8$ [Hz]
$f_{110} = (171.8) \sqrt{[(1/2)^2 + (1)^2 + (0)^2]} = 192.1$ [Hz] etc.

Larger rooms have lower modal frequencies. For example, for X = 200 [m], Y = 1 [m] and Z = 1[m],

$$f_{100} = (171.8) \cdot \sqrt{[(1/200)^2 + (0)^2 + (0)^2]} = 8.59 \text{ [Hz]}$$

a f_{100} ten times smaller than for X = 2 [m], Y = 1 [m] and Z = 1[m]

Given that for hard boundaries,

$$p_{lmn}(x, y, z, t) = A_{lmn} \cdot (\cos k_{xl} \cdot x) \cdot (\cos k_{ym} \cdot y) \cdot (\cos k_{zn} \cdot z) \cdot e^{j\omega t}$$
$$\text{for } l, m, n = 0,1,2,3,\ldots$$

and that for x = 0, y = 0 and z = 0 and for x = X, y = Y and z = Z

$$\cos k_{xl} \cdot x = 1$$
$$\cos k_{ym} \cdot y = 1$$
$$\cos k_{zn} \cdot z = 1$$

for all modes l. m, n = 0,1,2,3,..., the placement of sound sources and sound receivers at the corners of the room provide local maximum values for all modal components, while in any other point, x, y, z, in the room, some modal components are not at maximum, and some might be zero if in a node [21, 57].

Overall solution of the linear wave equation is obtained by superposition and is a triple infinite series (ignoring 0, 0, 0 mode corresponding to no sound transmission)

$$p(x, y, z, t) = \Sigma_{l=1 \text{ to } \infty} \Sigma_{m=1 \text{ to } \infty} \Sigma_{n=1 \text{ to } \infty}\ p_{lmn}(x, y, z, t)$$

or

$$p(x, y, z, t) = \Sigma_{l=1 \text{ to } \infty} \Sigma_{m=1 \text{ to } \infty} \Sigma_{n=1 \text{ to } \infty}\ A_{lmn} \cdot (\cos k_{xl} \cdot x) \cdot (\cos k_{ym} \cdot y) \cdot (\cos k_{zn} \cdot z) \cdot e^{j\omega t}$$

or,

$$p(x, y, z, t) = \Sigma_{l=1 \text{ to } \infty} \Sigma_{m=1 \text{ to } \infty} \Sigma_{n=1 \text{ to } \infty}\ A_{lmn} \cdot (\cos \pi \cdot l \cdot x / X) \cdot (\cos \pi \cdot m \cdot y / Y) \cdot (\cos \pi \cdot n \cdot z / Z) \cdot e^{j\omega t}$$

Example 6.2 FEMLAB™ Calculate of room eigenvalues assuming the simulated room dimensions X= 5 [m], Y= 4 [m] and Z = 2.6 [m], [56, 57].

The modal frequencies f_{lmn} [Hz] of the empty room with hard walls are given by

$$f_{lmn} = (171.8) \cdot \sqrt{[(l/5)^2 + (m/4)^2 + (n/2.6)^2]}$$
$$\text{for } l, m, n = 0,1,2,3,\ldots$$

such that the eigenvalues are

$$\lambda_{lmn} = \omega_{lmn}^2 = 343.6^2 \cdot \pi^2 [(l/5)^2 + (m/4)^2 + (n/2.6)^2]$$
$$\text{for } l, m, n = 0,1,2,3,\ldots$$

First 13 modal frequencies for l, m, n = 0, 1, 2 (000, 100, 010, 110, 001, 200, 101, 011, 210, 020, 111, 120, 201) are below the frequency of the mode 3,0,0 [37]

$$f_{300} = (171.8) \sqrt{[(3/5)^2 + (0/4)^2 + (0/2.6)^2]} = 103.08 \text{ [Hz]}$$

i.e. the eigenvalue

$$\lambda_{300} = (\omega_{300})^2 = (2 \cdot \pi \cdot f_{300})^2 = 4.19 \cdot 10^5 \text{ [rad}^2/\text{s}^2]$$

FEMLAB solution for hard boundary conditions for sound propagation in the room give for the first 15 eigenvalues up to around 10^5 [rad^2/s^2]. The eigenvalue $4.207857 \cdot 10^5$, computed by FEMLAB for the simulated room, is very close to the above computed eigenvalue for the empty room with hard walls and the same dimensions, $\lambda_{300} = 4.19 \cdot 10^5$.

Besides eigenvalues, FEMLAB permits to determine 3D pressure distribution in the simulated room for each eigenvalue.

The results for the sound pressure distribution are very different for various eigenvalues and are not used for the overall evaluation of room acoustic quality. Other modes have very different pressure distributions

[37]. As a consequence, sound pressure distributions are not directly useful in room acoustic design.

Other acousto-mechatronics applications can be found in [23, 58, 59, 60].

6.4 Experimental and Simulation Study of Room Acoustics

6.4.1 *Introduction*

Room acoustics is currently achieved more based on acoustic expert knowledge than on engineering design. Among the reasons for this situation are the limitations due to unsolved difficulties in solving inverse acoustic room problem given acoustic requirements. This section investigates means for experimental validation of simulation approaches for room acoustics design [88, 89].

Measurements of room acoustics require the development of a system that permits to retain from sensors outputs the part of the signal containing useful information and to remove the effect of measurement noise and of external noise. Sound dynamics in a room is modeled by partial differential equations for sound pressure or intensity. Such a model accounts for multiple wall reflections, refractions and attenuation. In this process, the sound in the room becomes increasingly complex, compared to the input signal and, moreover, the measured signals contain also acoustic measurement noise and noise transmitted from outside the room. Proper filtering of the measurement noise and outside noise benefits from the knowledge of the dynamic content of the acoustic signals received by microphones. For this purpose, in this section, room acoustics is numerically simulated to obtain a reference signal for the proper selection of the cutoff frequency of the low pass filters used to remove noise effects. As a result, useful content of the acoustic signal can be properly retained after signal filtering the experimental results. Moreover, the simulation model is also validated in this process and, as a result, this model can be used during the acoustic redesign of a room [88, 89].

6.4.2 *Proposed Approach*

For this acoustic investigation was chosen the rectangular room shown in Fig. 6.7, with a length of 9.4 [m], a width of 1.88 [m] and a height of 2.2 [m], subject external noise. This room, chosen for this preliminary simulation and experimental study, has concrete walls and contains no furniture.

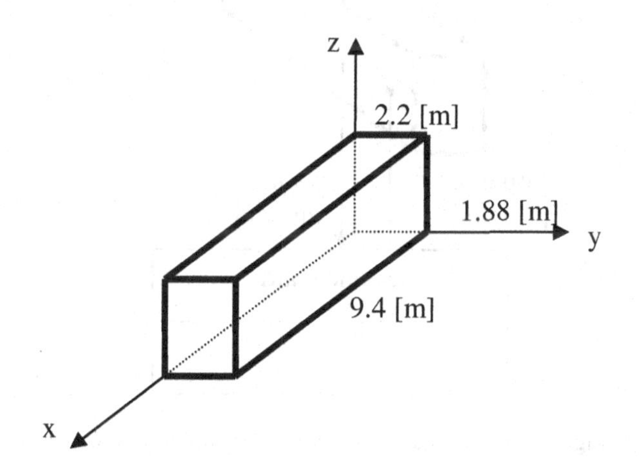

Fig. 6.7 Rectangular room for acoustic investigation

An experimental setup for room acoustics measurements is shown in Fig. 6.8. The location of the sound source (the speaker) is at about 2.1 [m] from the location of the receiver (the microphone). The surface reflection coefficients in this room with hard surfaces are approx. 0.85.

The diagram in Fig. 6.8 shows that, for a Single Input Single Output (SISO) case, a sound card is sufficient for providing analog voltage input to the speaker and for acquiring the analog voltage output from the microphone. Moreover, an advanced sound card has adequate sampling rate and resolution for an accurate room acoustic investigation. Microphone signals were acquired, saved, processed and displayed using MATLAB™ toolboxes. These signals were contaminated with outside and measurement noise and had to be signal conditioned. In particular, signal post-processing requires the proper choice of the cut-off frequency such that it leads to removing the noise while retaining the useful signal. This choice can be facilitated by a simulation study of room acoustics for the same input signal.

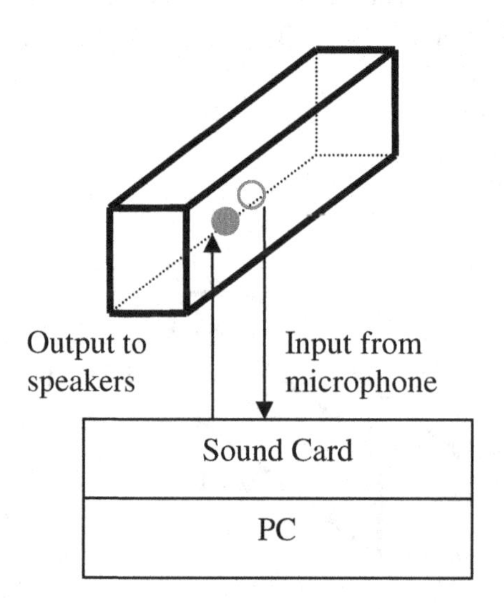

Fig. 6.8 Experimental setup for room acoustics measurements

6.4.3 *Simulation Model*

In order to simulate the transmission of the sound signal to the microphone location, following the direct propagation from the speakers and reflections from the wall, a room acoustics model, based on the impulse response of an enclosure, was developed using image method for acoustic ray propagation. Image method is one of methods in geometric room acoustics [82, 84, 86]. Allen and Berkley have developed an efficient method for obtaining impulse response of a rectangular room using image model technique [85]. In this case, a sound ray is used instead of a planar sound wave. The propagation of sound rays can be calculated using an image source to determine the length of the ray path. Assuming that the room is a homogeneous medium and that the refractions are negligible, the sound rays propagate in the straight lines in between walls. Other assumptions are: diffraction is negligible, the sound source is omni-directional and all walls have the same reflection coefficient. In the experimental study, in order to approximate an omni-directional speaker, an array of three speakers was used.

Preliminary simulations were carried out for an input given by a short pulse input signal, in order to evaluate if simulation model works as expected.

Simulation results for the signal intensity versus time at microphone location are shown in Fig. 6.9, for the case that rigid walls, and in Fig. 6.10, for non-rigid walls [88, 89].

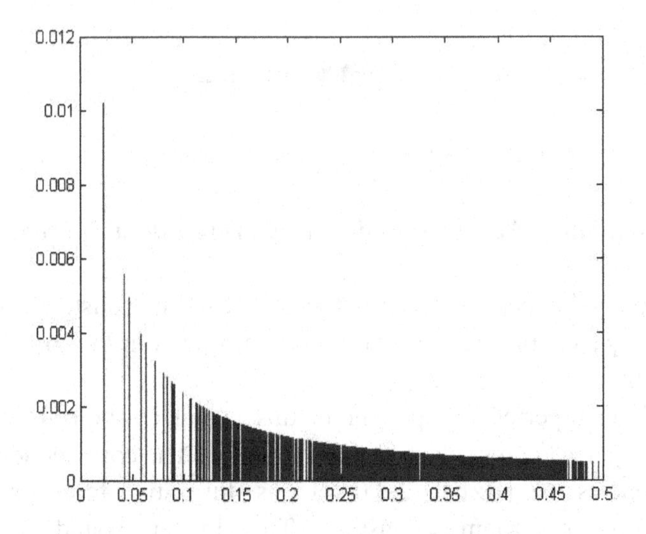

Fig. 6.9 The signal in the case of a room with rigid walls

The results show that the signals have decaying amplitudes and that the non-rigid walls lead to lower amplitudes versus time. These results are predictable and can be considered that they provide a first confirmation that the simulation model is correct. Complete confirmation will be the result of the subsequent simulation and experimental study for a more complex input signal.

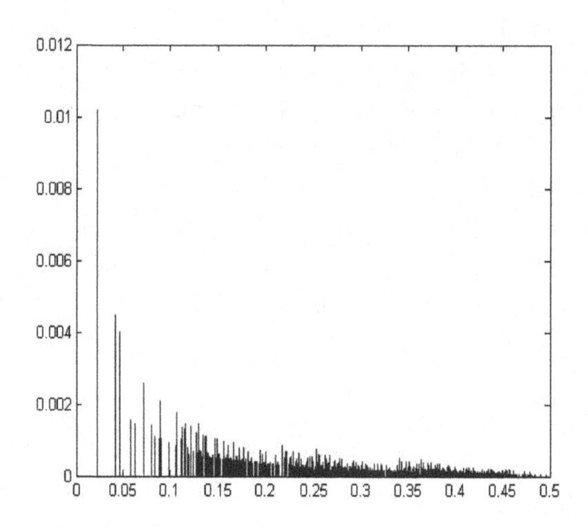

Fig. 6.10 The signal in the case of a room with non-rigid walls

6.4.4 *Simulation Results Based on Ray Propagation Approach*

Simulation and experimental investigation of room acoustics was carried out in this paper for a representative signal, shown in Fig. 6.11, selected from [87].

The longer period component in this signal is about 15 [ms] and the higher period component of about 0.5 [ms]. The corresponding range of frequencies, 67 [Hz] to 2000 [Hz] is illustrative for a preliminary investigation of room acoustics. The longer sound component wavelength is, in this case, approximately 2.5 m, and this limits the accuracy of the image method to geometric dimensions of the same order.

In fact, lowest modal frequencies for this room are 18 [Hz] along x, 90 [Hz] along y and 80 [Hz] along z, such that only along x there is a vibration mode with frequency lower than the input signal component with the frequency of 67 [Hz]. Consequently, it is expected that direct propagation and sound reflection along the length of the room shown in Fig. 6.8 are properly simulated and could serve as a reference for signal conditioning for the experimental study. Figure 6.12 shows the simulation results for the acoustic signal at microphone location, as a result of the output sound from the speaker, subject to the driving signal from Fig. 6.11 [88, 89].

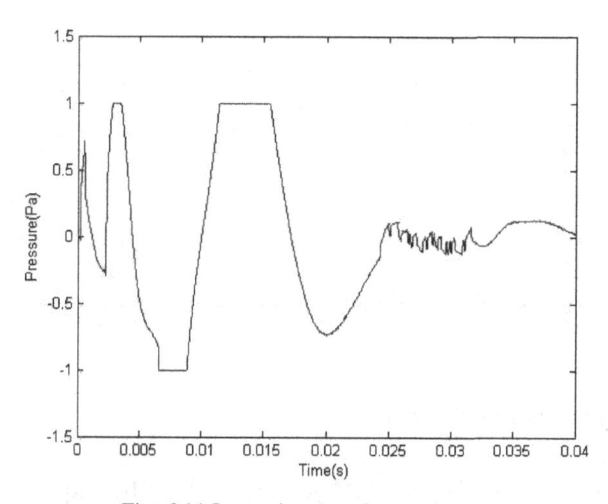

Fig. 6.11 Input signal to the speakers

Simulation results at the microphone location from Fig. 6.12 show no signal for the first 0.007 [s], *i.e.* for the time required to the direct sound to propagate over the distance of about 2.1 [m] from the speaker to the microphone. Afterwards, experimental results in fig. 6.12 from 0.007 to 0.014 [s] resemble the input signal from Fig. 6.11 for 0 to 0.007 [s] due to the fact that the delayed direct wave is received by the microphone during this time. Simulation results after 0.015 [s] differ from the input signal due to the effect of wall reflections. These signals differ more and more due to wall reflections. These signals will serve as a reference for the selection of the low-pass filter design for the experimental study.

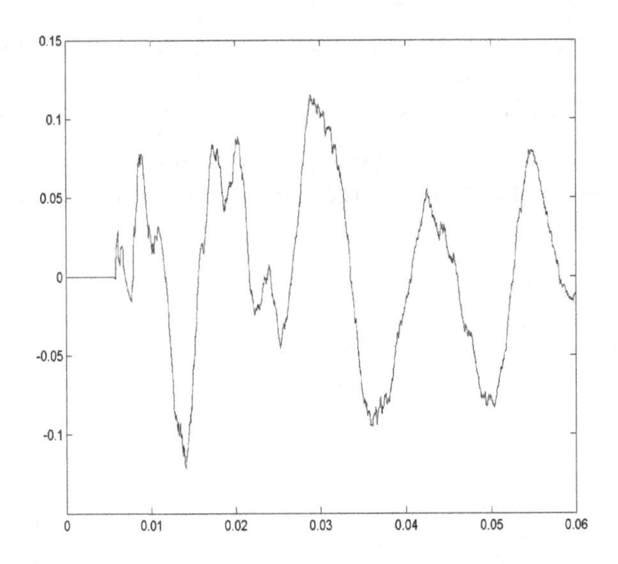

Fig. 6.12 Simulation results for the acoustic signal at microphone location

6.4.5 *Experimental Results*

The experimental study was carried out using the experimental setup shown in Fig. 6.8 and the input signal to the speaker shown in Fig. 6.11, also used for the above simulation study.

It was expected to have differences between the experiment results, for example, due to the fact that in experiments were used three speakers while in simulations was assumed an omni-directional sound source. Also, parametric uncertainty, for example in the case of reflection coefficients, can further lead to differences between experimental and simulation results. Moreover, measurement noise in microphone output further modify experimental results.

Figure 6.13 shows the non-filtered experimental result at microphone location [88, 89].

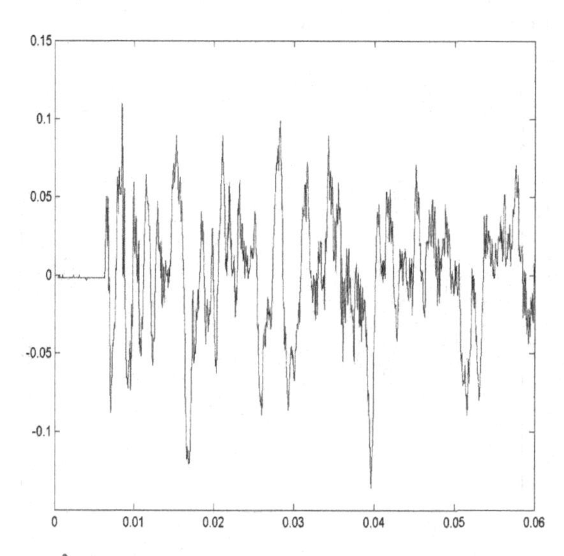

Fig. 6.13 Non-filtered experimental result at microphone location

The most obvious differences between the simulation result from Fig. 6.12 and the experimental result from Fig. 6.13 can be explained by the measurement noise in the experimental study. A significant reduction of these differences can be expected from the use of a low-pass filter to remove the noise [8]. Various values of the cut-off frequency were tried. The results are shown in Fig. 6.14 to 6.17, for the following cut-off frequencies: in Fig. 6.14, 2000 [Hz], in Fig. 6.15, 1000 [Hz], in Fig. 6.16, 800 [Hz], in Fig. 6.17, 500 [Hz] [88, 89].

The best agreement between simulation results from Fig. 6.12 and experimental results appears in Fig. 6.16 for a cutoff frequency of 800 [Hz]. Such agreement between experimental and simulation results also permits to conclude that the simulation model used in this investigation is validated, subject to satisfying the constraint that it is limited to geometric objects larger than the longest wavelength component of the input signal. Such a simulation model can, consequently be used for the acoustic redesign of the room under investigation, for example for the evaluation of the effects of changing wall materials, placing acoustic reflectors, furniture and audience etc.

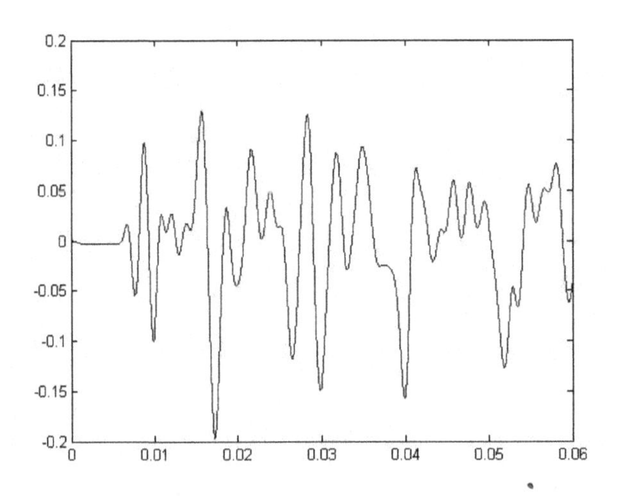

Fig. 6.14 The low-pass filtered signals for the cutoff frequency of 2000 [Hz]

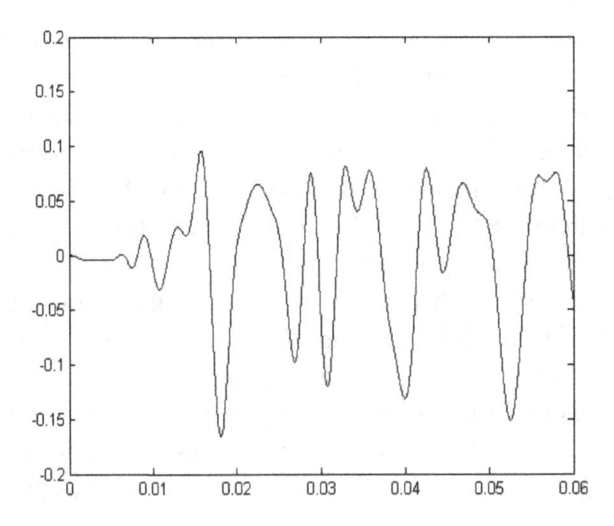

Fig. 6.15 The low-pass filtered signals for the cutoff frequency of 1000 [Hz]

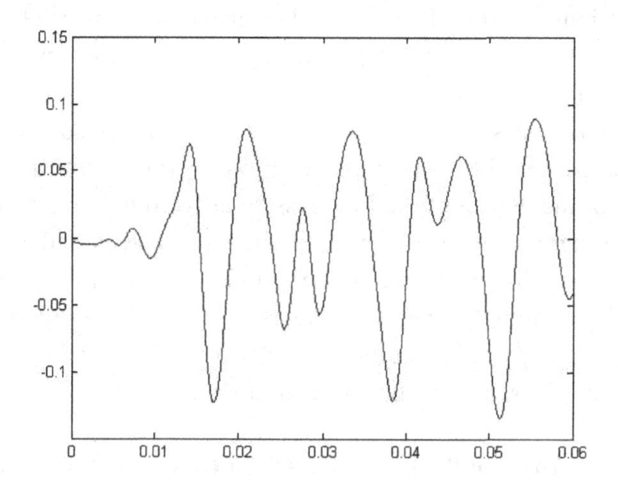

Fig. 6.16 The low-pass filtered signals for the cutoff frequency of 800 [Hz]

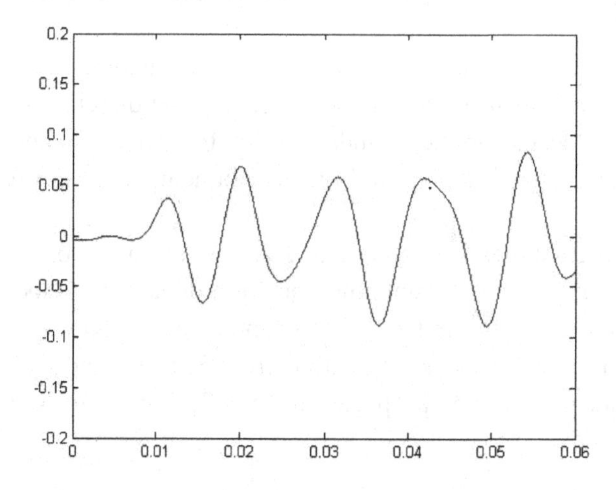

Fig. 6.17 The low-pass filtered signals for the cutoff frequency of 500 [Hz]

The room used in these experiments was a simple rectangular room with no furniture. More complex room geometry and the presence of furniture result in significantly richer sound signals and in such case the simulation study for the proper design of the low pass filter is significantly more useful.

This simulation study based on image method proved useful for the proper design of the measurement system for room acoustics, in particular for the design of the low-pass filter for removing measurement noise. The agreement between experimental and simulation results confirms the validity of the simulation model and justify its use acoustic redesign of the room. Further advantages of the proposed approach result from the possibility of assessing the efficiency of the location of the microphones for room acoustics measurement.

6.5 Discrete Inverse Problems Based on Direct and Reflected Ray Propagation

6.5.1 *Parameters Estimation Using Direct Ray Propagation*

Boundary measurement of direct ray propagation in distributed parameters systems use a linear model linking the direct wave reception by receptors located on the boundary, apart from one another, while the sources of the signal are located on the boundary opposite to emitters. [67].

Figure 6.20 shows I emitters and J receptors on the boundary of an acoustic field. Assume the unknown position dependent velocity v(x, y). The propagation from emitter E_i to receptor R_j, is characterized by the position variable along E_i to R_j, with α from E_i to R_j and the velocity is v(α) dependent of α. The propagation time T_{ij} from E_i to R_j is given by [67]

$$T_{ij} = \int_{E_i}^{R_j} \frac{d\alpha}{v(\alpha)}$$

In order to obtain a linear equation, v [m/s] in the denominator is replaced by slowness with s(α) unit [s/m],

$$s(\alpha) = 1 / v(\alpha)$$

such that

$$T_{ij} = \int_{E_i}^{R_j} s(\alpha) \cdot d\alpha$$

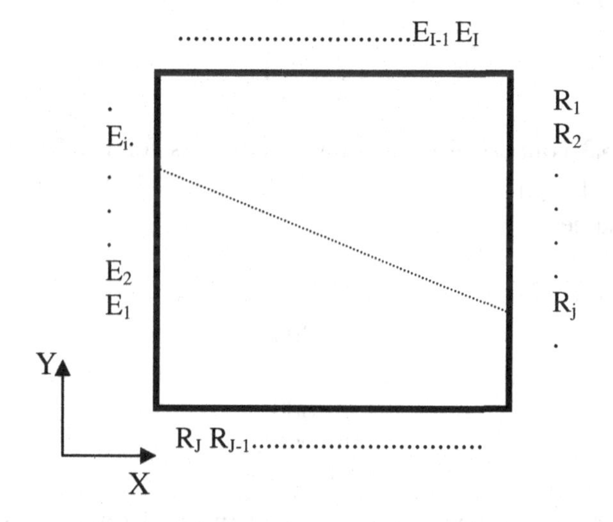

Fig. 6.20 Four pixel representation of direct ray propagation

The discrete form of the integration is obtained for K= pixels. For even numbers J = I, K is

$$K = (I / 2) \cdot (J / 2) = J^2 / 4$$

such that

$$T_j = \sum_{k=1}^{K} h_{ij} \cdot s_k \qquad \text{for j-1,2,\ldots, J}$$

where $h_{ij} = \alpha_k (i, j)$ is the length of the segment E_i to R_j, that crosses the pixel which has the slowness s_k. In matrix form, the direct problem is

$$\mathbf{T} = \mathbf{h} \cdot \mathbf{s}$$

where matrix **h** is [J · K].

The estimation of slowness s_k in each pixel k = 1,2,...,K, results form the inverse problem

$$\mathbf{s} = \mathbf{h}^{-1} \cdot \mathbf{T}$$

Example 6.3 Numerical estimation of slowness for J = I = 4 measurement values for y [1·4] from receptors R_j, j = 1, 2, 3, 4 and $K = J^2 / 4 = 4$ pixels. Assume

y =

2.1000
2.0000
2.1000
2.0000

and the four pixels shown in Fig. 6.21 of 1 [m] by 1 [m] each. The square matrix **h** [4 · 4] results as follows

h =

2 2 0 0
0 0 2 2
2 0 2 0
0 2 0 2

Is the problem even-determined?

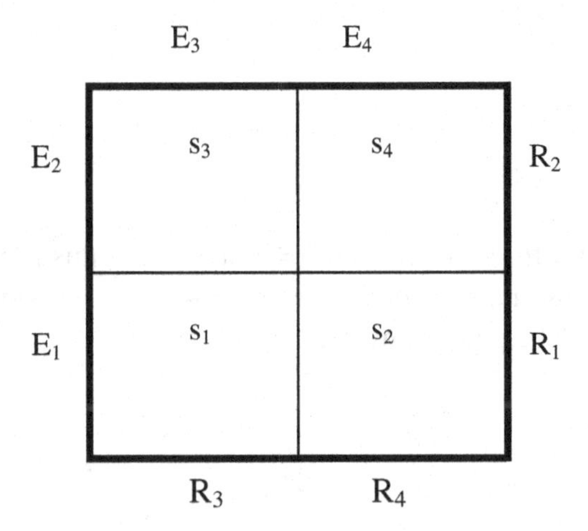

Fig. 6.20 Four pixel representation of direct ray propagation

MATLAB solution is

```
h=[2 2 0 0;0 0 2 2;2 0 2 0;0 2 0 2]

h =

    2   2   0   0
    0   0   2   2
    2   0   2   0
    0   2   0   2

>> inv(h)
Warning: Matrix is singular to working precision.

ans =

   Inf  Inf  Inf  Inf
   Inf  Inf  Inf  Inf
   Inf  Inf  Inf  Inf
   Inf  Inf  Inf  Inf
```

```
>> rank(h)

ans =

     3
```

The square matrix **h** [4 · 4] is of rank = 3 and makes this problem under-determined, for the number of unknowns = 4 and the solution requires the calculation of the pseudo-inverse.

```
>> pinv(h)

ans =

    0.1875   -0.0625    0.1875   -0.0625
    0.1875   -0.0625   -0.0625    0.1875
   -0.0625    0.1875    0.1875   -0.0625
   -0.0625    0.1875   -0.0625    0.1875

>> y=[2.1; 2; 2.1; 2]

y =

    2.1000
    2.0000
    2.1000
    2.0000

>> s=pinv(h)*y

s =

    0.5375
    0.5125
    0.5125
    0.4875

>>
```

Next example considers eight measurements y [8 · 1] for estimating four unknown slowness values s [4 · 1].

Example 6.4 Obtain the numerical estimation of slowness for an over-determined problem with

```
y =
      2.1000
      2.3000
      2.0000
      2.3000
      2.1000
      2.3000
      2.0000
      2.3000
```

and the non-square matrix **h** [8·4]

```
h =
    2.0000    2.0000         0         0
         0    2.3400    2.3400         0
         0         0    2.0000    2.0000
    2.3400         0         0    2.3400
    2.0000         0    2.0000         0
    2.3400         0         0    2.3400
         0    2.0000         0    2.0000
         0    2.3400    2.3400         0
```

MATLAB solution is

```
>> inv(h)
??? Error using ==> inv
Matrix must be square.

>> pinv(h)
```

```
ans =

0.1514  -0.0226  -0.0986   0.0843   0.1514   0.0843  -0.0986  -0.0226
0.1514   0.0843  -0.0986  -0.0226  -0.0986  -0.0226   0.1514   0.0843
-0.0986  0.0843   0.1514  -0.0226   0.1514  -0.0226  -0.0986   0.0843
-0.0986 -0.0226   0.1514   0.0843  -0.0986   0.0843   0.1514  -0.0226

>> rank(h)

ans =

   4
```

For

```
y =

   2.1000
   2.3000
   2.0000
   2.3000
```

the estimation of s results as follows

```
>> y=[2.1;2.3;2;2.3;2.1;2.3;2;2.3]

y =

   2.1000
   2.3000
   2.0000
   2.3000
   2.1000
   2.3000
   2.0000
   2.3000
```

```
>> s=pinv(h)*y

s =

    0.5253
    0.5003
    0.5003
    0.4753

>>
```

The pseudo-inverse is used for this over-determined problem, but input data have to be physically relevant and with reduced measurement noise to make the estimation accurate enough.

6.5.2 *Other Inverse Problems Using Ray Propagation*

Direct ray propagation model can also be used for the estimation of the unknown location of an emitter using a solution similar to triangularization using multiple receptors measurement of the signal arrival time form a single emitter. [67].

Boundary measurement of single ray reflection in distributed parameters systems is used in tomographic imaging [67]. It consists in an inverse problem of estimating the location of an anomaly given travel time measurements of signal from an emitter to the anomaly and reflected to a receptor. Reference [67] gives a detailed presentation and examples of application for crack location inside a solid body by measuring the first reflection time.

Problems

1. For a rectangular room, as shown in Fig. 6.6, with dimensions X = 9.4 [m], Y = 2.2 [m] and Z = 1.88 [m], calculate the natural frequencies for the first five modes, in case of sound speed in air c = 343.6 [m/s].

2. Estimate the slowness for J = I = 4 measurement values for y [1 · 4] from receptors R_j, j = 1 to 4 and K = 4 pixels, given

$$\mathbf{y} = \begin{matrix} 2.0100 \\ 2.0001 \\ 2.0500 \\ 2.0002 \end{matrix}$$

and **h** from Example 6.3.

3. Repeat problem 2 for

$$\mathbf{h} = \begin{matrix} 2 & 3 & 0 & 0 \\ 0 & 0 & 2 & 3 \\ 2 & 0 & 2 & 0 \\ 0 & 2 & 0 & 2 \end{matrix}$$

4. Obtain the numerical estimation of slowness for an over-determined problem with

$$\mathbf{y} = \begin{matrix} 2.0100 \\ 2.1000 \\ 2.0000 \\ 2.2000 \\ 2.1500 \\ 2.2000 \\ 2.0100 \\ 2.200 \end{matrix}$$

and **h** from Example 6.4.

Chapter 7

Thermo-Mechatronics

7.1 Direct Problem: Heat Flow Modeling and Simulation

7.1.1 *Direct Problem Solving for 2-Dimentional (2D) Heat Conduction from a Distributed Heat Source*

Direct problem refers to the effect of heat density or a distributed heat flux F(x, y, t) sources on the distributed parameters system temperature u(x, y, t). For 2-dimensional (2D) forced heat flow case the model is [25]

$$\frac{\partial u(x,y,t)}{\partial t} = k \cdot \left(\frac{\partial^2 u(x,y,t)}{\partial x^2} + \frac{\partial^2 u(x,y,t)}{\partial y^2}\right) + F(x,y,t)$$

or, in compact form

$$u' = k \cdot (u_{xx} + u_{yy}) + F$$

or, in a more compact form

$$u_t = k \cdot \nabla^2 u + F$$

where

u(x, y, t) is the temperature in a solid body in the point x, y at time t
k is diffusivity given by

$$k = K / (\sigma \cdot \tau)$$

σ is the specific heat of the solid body conducting the heat
τ is the volume density [kg/m^3]
F(x, y, t) is heat density (or heat flux) from an internal distributed heat source.

Boundary conditions are specific to each particular case. For example, for an insulated surface the boundary condition is given by $\nabla u \cdot n = 0$ where n is the vector normal to that surface.

Initial conditions take the form

$$u(x, y, 0) = \phi(x, y)$$

The method of separation of variables leads to the proposed solution

$$u(x, y, t) = X(x) \cdot Y(y) \cdot T(t).$$

Substituting it in the homogenous heat equation (F = 0)

$$u' = k \cdot (u_{xx} + u_{yy})$$

gives

$$X(x) \cdot Y(y) \cdot T'(t) = k \cdot (X''(x) \cdot Y(y) \cdot T(t) \text{ - } X(x) \cdot Y''(y) \cdot T(t))$$

or

$$\frac{T'(t)}{k \cdot T(t)} = \frac{X''(x)}{X(x)} + \frac{Y''(y)}{Y(y)}$$

To achieve validity for all x, y, t, this equation is separated into three ordinary differential equations, one first order and two second order

$$\frac{T'(t)}{k \cdot T(t)} = \lambda$$

$$\frac{X''(x)}{X(x)} = \lambda_1$$

$$\frac{Y''(y)}{Y(y)} = \lambda_2$$

where λ, λ_1 and λ_2 are constants to be determined separately.

The first ordinary differential equation, a first order equation, gives the time-dependent amplitudes. For the given initial conditions, Fourier series gives [25]

$$T_{mn}(t) = \frac{F_{mn}}{a_{mn}}(1 - e^{-a_{mn}t})$$

where the coefficients are given in specialized literature[25].

For the above boundary conditions, the solution u(x, y, t) of the direct problem for this linear system is obtained by superposition for the last two second order ordinary spatial differential equations as a double infinite series with sinusoidal shape functions

$$u(x, y, t) = \sum_{n=1}^{\infty} \sum_{m=1}^{\infty} T_{mn}(t) \cdot \sin(n \cdot \pi \cdot x) \cdot P_{mn} \cdot \sin(m \cdot \pi \cdot y)$$

This solution verifies the above boundary conditions for $\lambda_1 = n \cdot \pi$ and $\lambda_2 = m \cdot \pi$ for n, m = 0, 1, 2, 3....

The general solution becomes

$$u(x, y, t) = \sum_{n=1}^{\infty} \sum_{m=1}^{\infty} (\frac{F_{mn}}{a_{mn}}) \cdot (1 - e^{a_{mn}t}) \cdot \sin(n \cdot \pi \cdot x) \cdot P_{mn} \cdot \sin(m \cdot \pi \cdot y)$$

This direct problem solution is not in closed form and, as in Ch. 6, this results in major difficulties in solving the inverse problem of determining F(x, y, t) for achieving a desired u(x, y, t).

MAPLE and FEMLAB examples illustrate the use of double infinite series solution of the heat flow direct problem.

7.1.2 *Direct Problem Simulation of 2D Heat Flow for a Continuous Point-Heat Source Input Using MAPLE™*

As an example, the following 2-D heat flow problem, for a square thin plate of 1 [m] by 1[m], with the temperature fixed at 0 at all points along the square boundaries, is presented in the form of MAPLE plot, shown in Fig. 7.1, for a particular constant internal heat source, given by [25, 107]

$$F(x, y) = 30 \cdot \sin(2 \cdot \pi \cdot x) \cdot \sin(2 \cdot \pi \cdot y) \quad \text{for } 0 \leq x \leq 0.5 \text{ and } 0 \leq y \leq 0.5$$
$$= 0 \quad \text{for } 0.5 < x \leq 1 \text{ and } 0.5 < y \leq 1$$

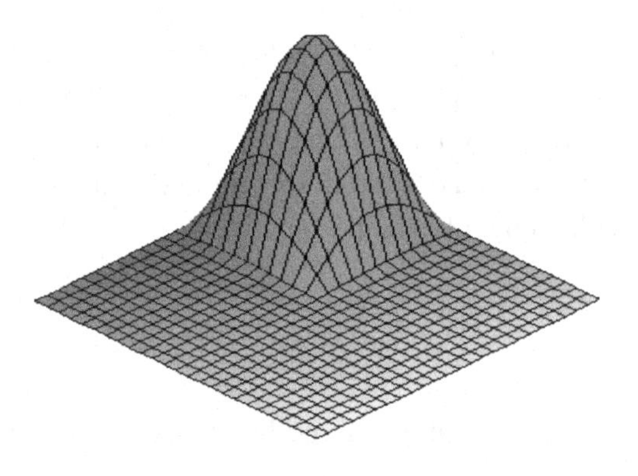

Fig. 7.1 MAPLE plot for the internal heat source

The results are based on the above double series solution of the heat flow equation, for the non-oscillatory time variation of the temperature u(0.5, 0.5, t) at the heat source located in the center (0.5. 0.5) of the square plate.

Results for the spatial variation of the temperature u(x, y, 0.05) are shown in Fig. 7.2.

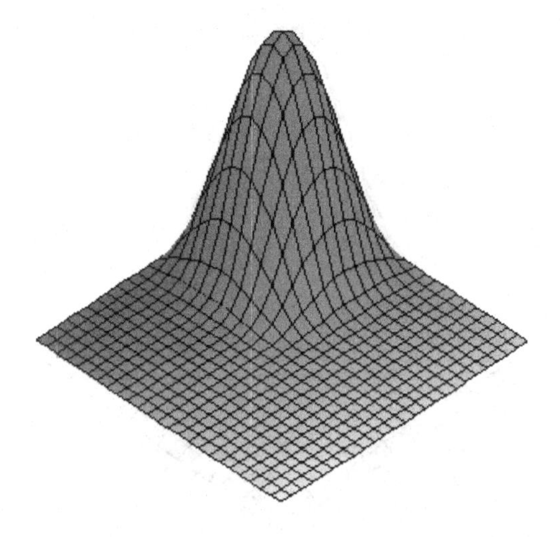

Fig. 7.2 Spatial variation of the temperature u(x, y, 0.05)

Results for the spatial variation of the steady state temperature u(x, y, ∞) are shown in Fig. 7.3. Spatial temperature distribution over time, shown in Fig. 7.1 for t = 0, Fig. 7.2 for t = 0,5 [s] and Fig. 7.3 for $t \rightarrow \infty$, is dominated by a uni-modal shape with maximum at (0.25, 0.25), which coincides with the location of the maximum of heat source temperature, shown in Fig. 7.1.

These results show that single input F(x, y) cannot generate an arbitrarily shaped desired temperature distribution and also that temperature sensors, located at the plate boundaries, receive delayed and reduced temperature variations over time. This illustrates the difficulties in solving inverse problems of the control of point sources of heat to achieve a desired temperature distribution and of remote monitoring of the temperature. In fact, this remote monitoring of temperature was one of the first studied ill-posed inverse problems [30].

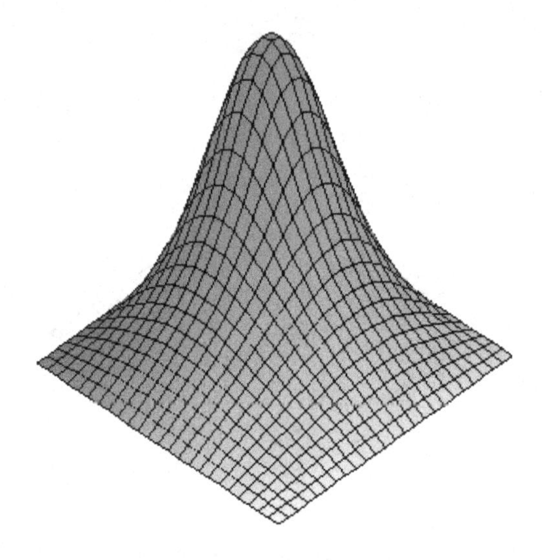

Fig. 7.3 Spatial variation of the steady state temperature u(x, y, ∞)

7.1.3 *Direct Problem Simulation of 2D Heat Flow for a Short Temperature Pulse Input Using FEMLAB™*

In this section, FEMLAB™ simulations are carried out for an unknown heat source located at the center of a rectangular plate that produces a fast temperature change, represented as a short pulse [56, 57]. The simulations were carried out for a rectangular steel plate of dimensions 1.5 [m] by 2 [m], subject to a heat source of 0.2 [m] diameter at the center. Outer boundary is assumed thermally insulated at all sides. Inner domain of the circular heat source has boundary conditions that correspond to a very intense heat source $Q = 10^6$ MW for $t < 1$ (*i.e.* for a heat source that is on for less than 1 second and then is shut-off.). Initial temperature is 0 [^{0}C].

Figures 7.4–7.9 show the 2D spatial distribution of the temperature u(x, y, t) versus time [36].

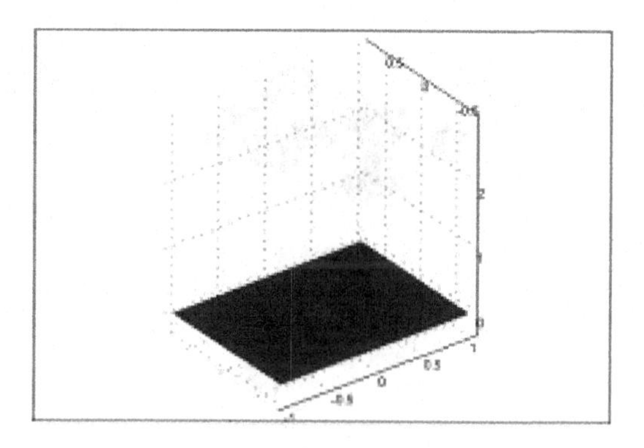

Fig. 7.4 Temperature u(x, y, 0.0)

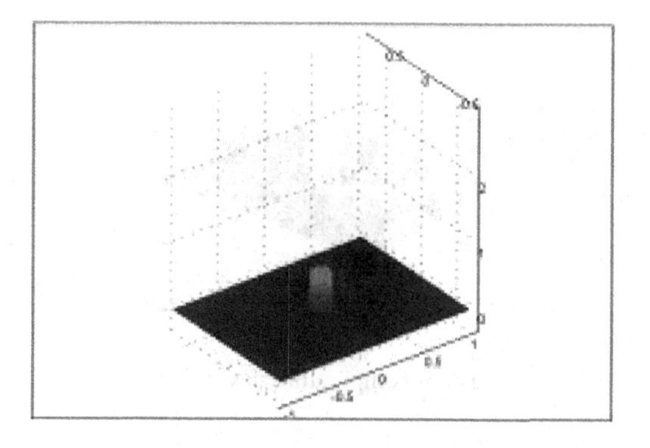

Fig. 7.5 Temperature u(x, y, 0.2)

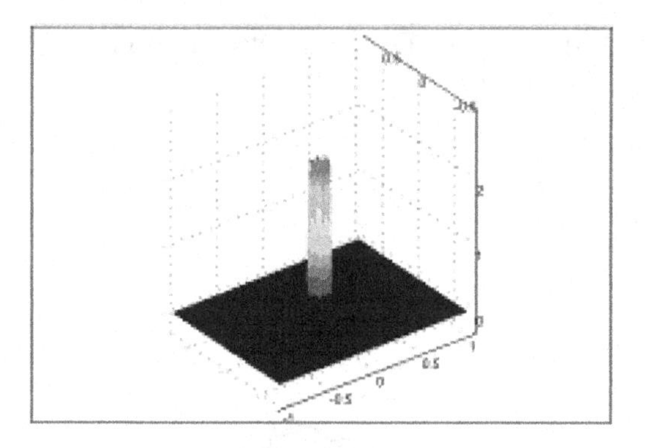

Fig. 7.6 Temperature u(x, y, 0.8)

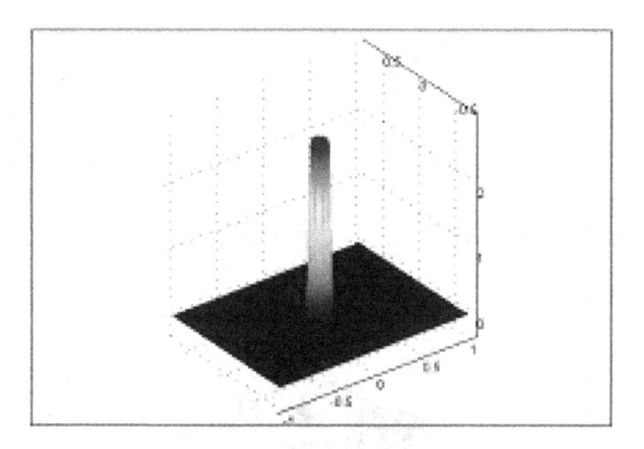

Fig. 7.7 Temperature u(x, y, 5.0)

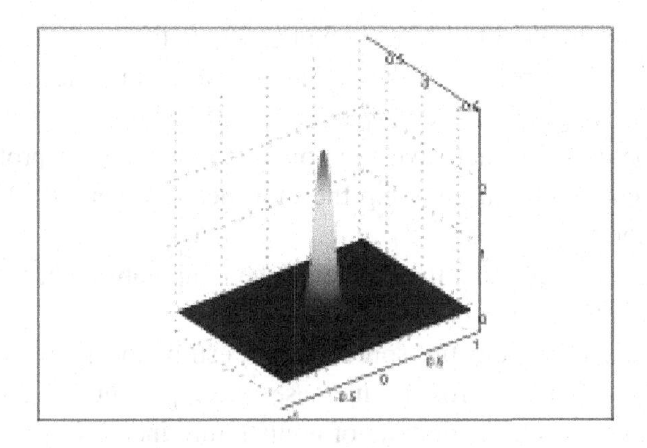

Fig. 7.8 Temperature u(x, y, 80)

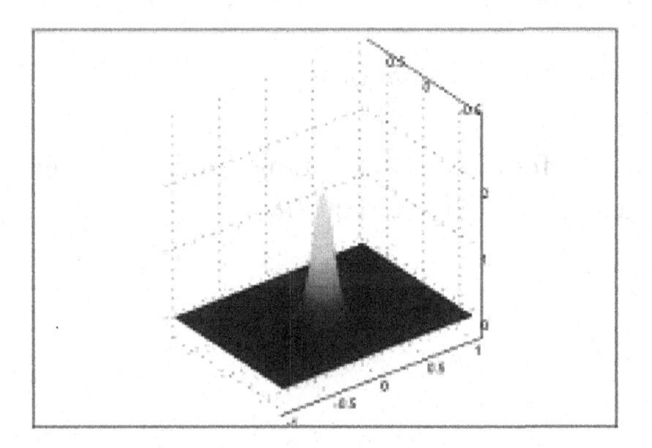

Fig. 7.9 Temperature u(x, y, 175)

It can be observed that the fast and significant rise in time of the temperature at the location of the heat source, in the center of the rectangular plate, leads to slow and a less significant temperature increase at locations further away from the heat source. These locations are where sensors were assumed to have been placed. The inverse problem difficulties in reconstituting (estimating) the heat source input into the center of the plate from remote temperature sensing are clearly illustrated by these results. These difficulties arise in general in the case of remote sensing, when the propagation of the signal leads to ill-posed inverse problems. Further investigations for solving such problems can rely on recent results in applying the mollification method [22, 47] and regularization methods [41, 93].

It can be observed that the values of the temperature decreases as the distance from the internal heat source increases.

As the distance from the source to the estimation point increases it takes much longer time for the increase in temperature to arrive at the boundary and, moreover, the rate of temperature increase is significantly lower.

These results indicate that significant analytical and simulation studies are required for the investigation and testing of solutions to the inverse heat problem to be solved for remote monitoring.

7.1.4 *Direct Problem Formulation for 3-D Heat Flow*

Direct problem is formulated for 3-dimensional (3D) heat conduction problem by the non-homogenous equation

$$u_t = k \cdot \nabla^2 u + F$$

or

$$\frac{\partial u}{\partial t} = k \cdot (u_{xx} + u_{yy} + u_{zz}) + F$$

where

$u(x, y, z, t)$ is the temperature in a solid body in the point x, y, z at time t.
k is diffusivity given by
$k = K / (\sigma \cdot \tau)$
σ is the specific heat of the solid body conducting the heat
τ is the volume density [kg/m^3]

$F(x, y, z, t)$ is heat density (or heat flux) from an internal heat source.

Issues in remote monitoring and control with point sensors and point actuators are the same as for the above 2D example, but obviously are even more difficult to solve in real-time.

7.2 Inverse Problem Solution for Remote Temperature Monitoring

7.2.1 *Introduction*

In this section remote sensing issues will be analyzed taking into account that monitoring in this case can lead to inverse problems that are ill-posed. For this case of sensing the solutions of the resulting ill-posed problem of estimation of internal variables of a system from measurements on the boundary of the system will be investigated.

These difficulties are typical issues in real-time remote temperature sensing [30, 36].

Advanced systems require sensing, acquisition and processing of signals from multiple sensors [31].

The development of such systems cannot rely only on traditional design tools and requires extensive investigations of new tools based, among others, on non-linear discontinuous systems modeling of the field propagation of signals and power from the input throughout the system and to the sensors.

In this section the focus is on real-time remote temperature sensing, away from heat source location.

Temperature measurement and heat flow estimation has been analyzed for other distributed parameters applications and the inverse heat problem solving difficulties were identified and investigated [42, 43, 35]. For example, for the case of an explosion sensing, the consequences and the solutions of the resulting ill-posed problem of estimation of internal variables of a system from measurements on the boundary of the system are investigated in [22, 30, 44]. Local or distributed catastrophic events are spreading in seconds to hours and the propagation medium operates as a low pass filter that filter out useful high frequency signal components before reaching remote sensors. This physical low pass filter effect is the main cause for inverse problems to become ill-posed.

In field propagation of signals, three cases can be considered:

A) Locally initiated heat source (for example, explosions producing heat and/or acoustic waves that spread in the free space, floods *etc.*).
B) Distributed and moving sources (hurricanes, air and water biological contamination, etc).
C) Infrared waves associated with the heat generation that propagate faster over longer distances in free space and with less significant attenuation such that their propagation medium acts to a much lesser extend as low pass filters.

Remote temperature sensing is more efficient with infrared radiation sensors, but this is not possible in the case of solid or liquid propagation fields.

7.2.2 *Inverse Problem for Heat Flux Input Remote Estimation from Temperature Measurements*

This section presents the analytical solution for the identification of the difficulties in remote estimation of the unknown heat flux input based on the output from sensors located on given boundaries of the thermal field. Inverse heat conduction problem is first analyzed considering straight line 1D heat conduction equation for $x > 0$ and $t > 0$

$$\frac{\partial^2 u(x,t)}{\partial x^2} = \frac{\partial u(x,t)}{\partial t}$$

where $u(x, t)$ is the temperature (in dimensionless units) in point x at time t.

Exact measurements of the temperature $u(x_m, t)$ come from the temperature output $T_m(t)$, of the sensor located at $x = x_m$

$$u(x_m, t) = T_m(t)$$

Boundary conditions are:

$u(x, t) < \infty$ for a semi-infinite body
$u(0, t) = T(t)$ in case that the temperature at x = 0) is the unknown
$u_x(0, t) = - q(t)$ in case of unknown surface heat flux q(t) entering at x = 0.

In Ch. 2.3 Fourier transform approach was used for the study of frequency effect in the case of the non-collocated temperature input.

The equation

$$q = -k \cdot \nabla u$$

at the surface (x = 0), q(0, t) can also be solved using infinite series approach proposed by Burggraf [30].

The time-domain solution of the 1D equation for remote estimation of heat flux $q_{est}(0, t)$ from temperature measurements T, is obtained using infinite series approach as follows [25, 30]

$$q_{est}(0, t) = q(x_m, t) + \sum_{n=1}^{\infty}\{x_m^{2n-1} / [k^n \cdot (2n + 1)!]\} \cdot (d^nT / dt^n) + \sum_{n=1}^{\infty}\{x_m^{2n} / [k^n \cdot (2n)!]\} \cdot (d^n q(x_m, t) / dt^n)$$

where the heat flux $q(x_m, t)$ and its time derivatives are calculated using a similar procedure based on infinite series approach.

It can be observed that the higher the derivatives d^nT / dt^n the larger the values of the multiplicative coefficients $\{x_m^{2n-1}/[k^n \cdot (2n+1)!]\}$ and $\{x_m^{2n} / [k^n(2n)!]\}$ for larger distances between sensor location $x = x_m$ and the surface x = 0. This exact result consists in a series expansion for higher derivatives d^nT / dt^n and corresponds to the multiplicative coefficient function of frequency ω from Ch. 2.3

$$\exp\{-\sqrt{(|\omega| / 2)}[1+ i \cdot \mathrm{sgn}(\omega)]\}$$

Similar difficulties appear in the case of other inverse problems for distributed parameters systems described by elliptic PDE, i.e. for other cases of using acoustic or vibration sensing [34, 52, 44].

The above equation for the estimation of the heat flux $q_{est}(0, t)$ taking into account the effect of the distance x_m from the heat flux input to the location of the sensor shows that the estimation for higher frequency components, that characterizes fast varying signals, becomes more and more difficult as the distance x_m increases.

Problems

1. Assume the 2-D heat flow problem, for a square thin plate of 2 [m] by 3[m], with the temperature fixed at 0 at all points along the square boundaries.

   Obtain a MAPLE plot, for a constant internal heat source, given by

$$F(x, y) = 20 \cdot (\sin 2 \cdot \pi \cdot x) \cdot (\sin 2 \cdot \pi \cdot y) \quad \text{for } 0 \leq x \leq 1 \text{ and } 0 \leq y \leq 1$$
$$= 0 \quad \text{for } 1 < x \leq 2 \text{ and } 1 < y \leq 3$$

2. Simulate the heat flow in a rectangular steel plate of dimensions 1 [m] by 3 [m], subject to a heat source of 0.1 [m] diameter at the center. Outer boundary is assumed thermally insulated at all sides. Inner domain of the circular heat source has boundary conditions that correspond to a very intense heat source $Q = 10^8$ [MW] for a heat source that is on for less than 1 second and then is shutoff. Initial temperature is 0 [^{0}C].

Chapter 8

Magneto-Mechatronics

8.1 Introduction

The continuously increasing demands for fast and accurate position control systems from opto-electronics, computer hardware and peripherals, precision machining, robotics and, recently, auto industry, have stimulated the interest in the use of new, non-conventional implementations of position control. An interesting solution is offered by the use of magnetic bearings for avoiding dry friction, a major source of reduced precision in positioning.

This section will review recent results regarding magnetic bearing models, observers and controllers. Models and observers for magnetic levitation, magnetic bearings and combined frictionless motor-bearing systems will first be presented. The second part will be dedicated to the presentation of magnetic bearings and motor-bearings systems controllers (PID, state feedback, LQ, time-delay, PDD, PIDD, feedback linearization, state derivative feedback and integral controllers etc.). Simulation examples will illustrate the models and the controllers.

In the presentation, a single axis nonlinear model is used. This model permits to focus on the performance of the proposed controllers, using SIMULINK™ simulators for a linear (PID) controller and then for nonlinear controllers (feedback linearization based controller and for a state derivative feedback controller) that will illustrate the relative improvements in case of nonlinear controllers compared to a linear controller [94].

Feed-forward compensation controllers, as well as feedback linearization with "integral control" and a full order observer were tested on 1D magnetic levitation models [72].

The results of the effect of sampling rate taking into account fixed time delays, analog low pass-filter, and zero-order-hold device effects on arious linear (PD, PDD and PIDD) controllers are documented in [95].

Approaches based on time delay control are presented in [96-98]).

State feedback for azimuth motion of a frictionless positioning device is developed in [97]. Magnetic bearing control used one-step-backward plus current step signals for nonlinear compensation.

LQ control of magnetic bearings, using 1 DOF and 5 DOF models for magnetic bearings of a brushless motor was developed in [99]. Hysteresis in electromagnetic actuators can be compensated through Preisach model inversion. The goal was the compensation of hysteresis effects in magnetic suspension systems using soft ferromagnetic materials. A 1 DOF model of the system was used [101].

In a recent review, of industrial solutions for magnetic bearing systems control, only linear controllers were listed [104].

8.2 Direct Model

The model considered in this section for the control of magnetic suspension systems through nonlinear control schemes is that of a single axis system used for maintaining a ball at a desired height when it is subjected to external disturbances [72, 105, 106]. Multi axis systems require complex models that would not serve, at this stage of the analysis, the purpose of comparing the performance improvement of nonlinear controllers versus linear controllers. Fig. 8.1 shows a single axis magnetic levitation system used in this section for control analysis.

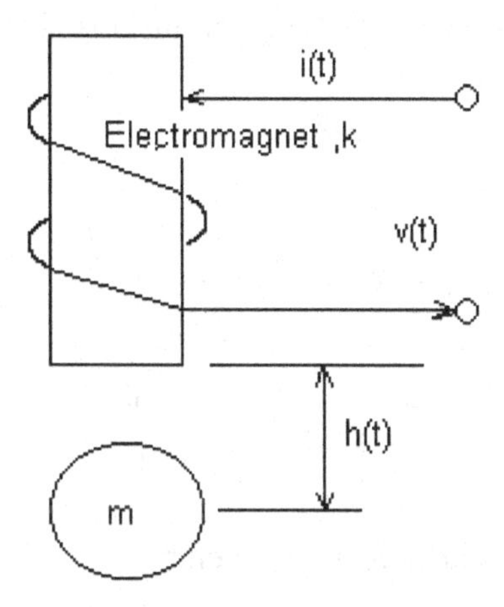

Fig. 8.1 Single axis magnetic suspension system

The system shown in Fig. 8.1 can be modeled by the 1 DOF (Degree of Freedom) nonlinear equation of motion [72, 127] and the voltage equation for the electric circuit

$$m\frac{d^2h(t)}{dt^2} = m \cdot g - k \cdot (i(t) / h(t))^2$$

$$L\, di(t)/dt = v(t) - Ri(t)$$

The general form of an affine system

$$dx / dt = f(x) + g(x) \cdot u$$

is obtained by denoting variables for state space representation as follows

$$x_1 = h$$
$$x_2 = dh / dt$$
$$x_3 = i$$
$$u = v = R \cdot i + L \cdot di / dt$$

such that the scalar equations of the nonlinear state space model are

$$\dot{x}_1 = x_2$$

$$\dot{x}_2 = -\frac{k}{m} \cdot \frac{x_3^2}{x_1^2} + g$$

$$\dot{x}_3 = -\frac{R}{L} \cdot x_3 + \frac{1}{L} \cdot u$$

Nonlinear model in matrix form is given by

$$\begin{bmatrix} \frac{dx_1}{dt} \\ \frac{dx_2}{dt} \\ \frac{dx_3}{dt} \end{bmatrix} = \begin{bmatrix} x_2 \\ \frac{-k \cdot x_3^2}{m \cdot x_1^2} + g \\ -(R/L) \cdot x_3 \end{bmatrix} + \begin{bmatrix} 0 \\ 0 \\ 1/L \end{bmatrix} \cdot u$$

For R = 0 and L = 1, the equilibrium position h_0 is maintained by a current

$$i_0 = h_0 \cdot \sqrt{(m \cdot g / k)}$$

For m = 1 [Kg], g = 9.81 [m/s^2] and k = 0.1 [Nm^2/A^2] the equilibrium position for $h_0 = 0.02$ [m] is maintained by $i_0 = 0.198$ [A].

8.3 Simulation Results for Linear Control

The PID controller shown in Fig. 8.2 has as input the error (x_d - x_1) and generates the command u for the nonlinear model given above in matrix form. The output of the "Nonlinear model" block is the position variable $x_1 =$ h [107].

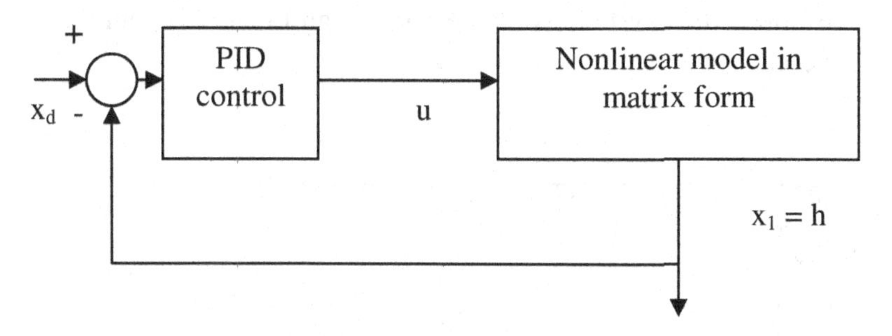

Fig. 8.2 PID control

The PID controller is given in the SIMULINK model by the s-Function

```
[num,den] = tfdata(150*tf(-1*[1 20],[1 50])*tf([1 1],[1 0]),'v')
num = [-150 -3150 -3000]
den = [1 50 0]
```

where the proportional gain is chosen 150 and the PD control gains are chosen from root locus analysis with a zero at 20 and a pole at 50. The integral control is given by (s + 1) / s [105].

The system is controlled for a specific desired value x_d. Feed-forward compensation of gravity effect $m \cdot g$ is achieved by applying an input voltage $v_0 = R \cdot i_0 = R \cdot h_0 \cdot \sqrt{(m \cdot g / k)}$.

For R = 5, m = 1, g = 9.81 and k = 0.1, the equilibrium position for $h_0 = 0.02$ is maintained by $v_0 = R \cdot i_0 = 5 \cdot 0.1982 = 0.991$ [V[. This input voltage permits to keep the magnetic ball in the equilibrium position h_0 [94].

Simulations were carried out for

a) step input of amplitude $x_d = 0.018$ [m] and the initial condition h(0) = 0.03 [m]
b) step input of amplitude $x_d = 0.03$ [m] and the initial condition h(0) = 0.18 [m].

a) Simulation results for step input of amplitude $x_d = 0.018$ [m] and the initial condition h(0) = 0.03 [m] are shown in Fig. 8.3. As seen from the response, the system is stable for a significant off-equilibrium initial condition.

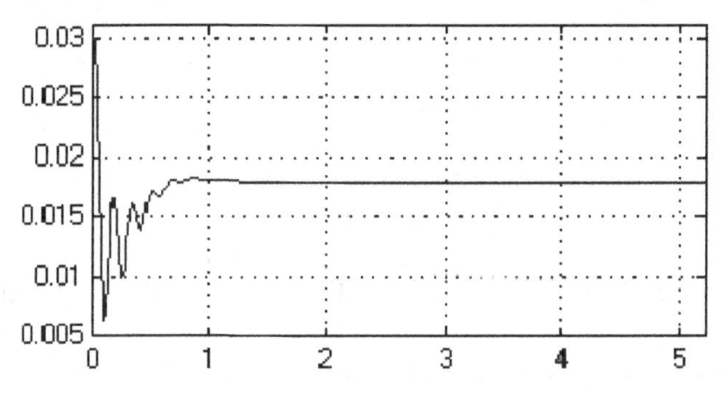

Fig. 8.3 Position h(t) for a step input of amplitude $x_d = 0.018$ [m] and initial position h(0) = 0.03 [m]

b) A higher value of the step input makes the system become unstable and the ball falls outside the region where the attraction force of the electromagnet is effective. Fig. 8.4 shows the simulation results for step input of x_d =0.03 [m] and initial position h(0) = 0.018 [m] and indicates an unstable system with h(t) diverging indefinitely.

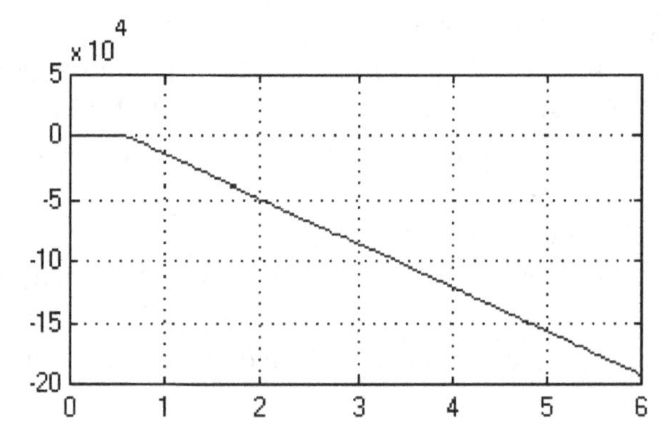

Fig. 8.4 Position h(t) for a step input of amplitude $x_d = 0.03$ [m] and initial position h(0) = 0.018 [m]

8.4 State-Input Linearization of a Magnetic Levitation System

8.4.1 *Feedback Linearization*

Feedback linearization approach defines new state variables function of x_1, x_2 and x_3 and a new control variable v generated by a linear feedback controller using a nonlinear input transformation.

For an affine system, a state transformation z =T(x) results from the conditions [74]

$$\frac{\delta T_1}{\delta x_1} \neq 0$$

$$\frac{\delta T_1}{\delta x_2} = 0$$

$$\frac{\delta T_1}{\delta x_3} = 0$$

Choosing a simple solution

$$z_1 = T_1 = x_1$$

gives

$$z_2 = T_2 = \frac{\delta T_1}{\delta x} f = \begin{bmatrix} \frac{\delta T_1}{\delta x_1} & \frac{\delta T_1}{\delta x_2} & \frac{\delta T_1}{\delta x_3} \end{bmatrix} \cdot \begin{bmatrix} x_2 \\ \frac{-k \cdot x_3^2}{m \cdot x_1^2} + g \\ 0 \end{bmatrix} = x_2$$

$$z_3 = T_3 = \frac{\delta T_2}{\delta x} f = \begin{bmatrix} 0 & 1 & 0 \end{bmatrix} \cdot \begin{bmatrix} x_2 \\ \frac{-k \cdot x_3^2}{m \cdot x_1^2} + g \\ 0 \end{bmatrix} = -\frac{k \cdot x_3^2}{m \cdot x_1^2} + g$$

The resulting State Transformation $z = T(x)$ is [72]

$$z_1 = x_1$$
$$z_2 = x_2$$
$$z_3 = -\frac{k \cdot x_3^2}{m \cdot x_1^2} + g$$

Time derivative of z_3 gives

$$\dot{z}_3 = -2 \cdot \frac{k}{m}\left[\frac{x_3}{x_1^2}\dot{x}_3 - \frac{x_3^2}{x_1^3}\dot{x}_1\right]$$

Inverse state transformation $x = T^{-1}(z)$ is

$$x_1 = z_1$$
$$x_2 = z_2$$
$$x_3 = z_1\sqrt{(g - z_3)\frac{m}{k}}$$

The original state space model was

$$\dot{x}_1 = x_2$$

$$\dot{x}_2 = -\frac{k}{m} \cdot \frac{x_3^2}{x_1^2} + g$$

$$\dot{x}_3 = -\frac{R}{L} \cdot x_3 + \frac{u}{L}$$

A new control variable v is defined as

$$v = \dot{z}_3$$

such that

$$\dot{z}_3 = -2 \cdot \frac{k}{m}[\frac{x_3}{x_1^2}\dot{x}_3 - \frac{x_3^2}{x_1^3}\dot{x}_1]$$

becomes

$$v = -2 \cdot \frac{k}{m}[\frac{x_3}{x_1^2}(-\frac{R}{L} \cdot x_3 + \frac{u}{L}) - \frac{x_3^2}{x_1^3} x_2]$$

The solution for u gives

$$u = -\frac{m \cdot L}{2 \cdot k}\frac{x_1^2}{x_3} \cdot v + (R + L \cdot \frac{x_2}{x_1}) \cdot x_3$$

For the linearized system, a full state feedback control law can now be applied. The original nonlinear state space model with states x_1, x_2 and x_3 and input u, subject to the above nonlinear control u and state transformation $z = T(x)$, results in a linearized system with new states z_1, z_2 and z_3 and new input v [127]. A Linear Full State feedback can generate the new control variable, v.

8.4.2 *State-Input Linearization and Linear Feedback Control*

The new control variable, v, is given by the Linear Full State feedback equation

$$v = - G_1 \cdot (z_1 - x_d) - G_2 \cdot z_2 - G_3 \cdot z_3$$

Figure 8.5 shows the block diagram of the feedback linearization control, used in simulations for m = 1.

The blocks from Fig. 8.5 are the following:

-Linear Full State Feedback to generate v given z_1, z_2 and z_3
-Nonlinear Input Transformation of v into u
-Nonlinear System Model for the states x_1, x_2 and x_3 given input u
-State Transformation z = T(x).

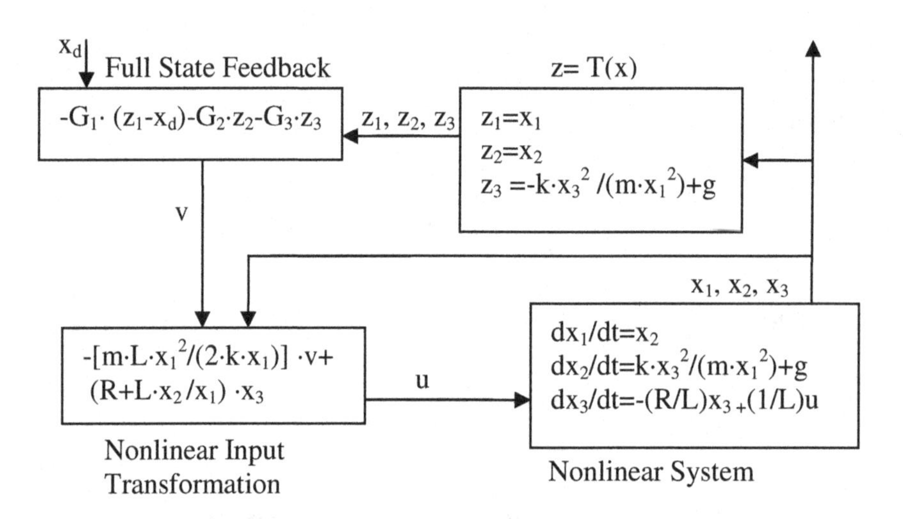

Fig. 8.5 Full state feedback control and feedback linearization of the nonlinear system

A SIMULINK program was designed based on the block diagram from Fig. 8.5. Simulations were carried out for a step input of $x_d = 0.035$ [m] an initial position h(0) = 0.022 [m] and gains $G_1 = 32$, $G_2 = 32$ and $G_3 = 0$ [94]. The results are shown in Fig. 8.6

Feedback linearization and full state feedback based controller lead to a stable response in the case in which a PID control resulted in an unstable system. The desired position $x_d = 0.035$ [m] for the case shown in Fig. 8.6 is far from the equilibrium position of 0.02 [m] and is reached after a large overshoot. The much smaller overshoot from Fig. 8.3 is explained by a desired position of 0.018 [m], much closer to the equilibrium position of 0.02 [m].

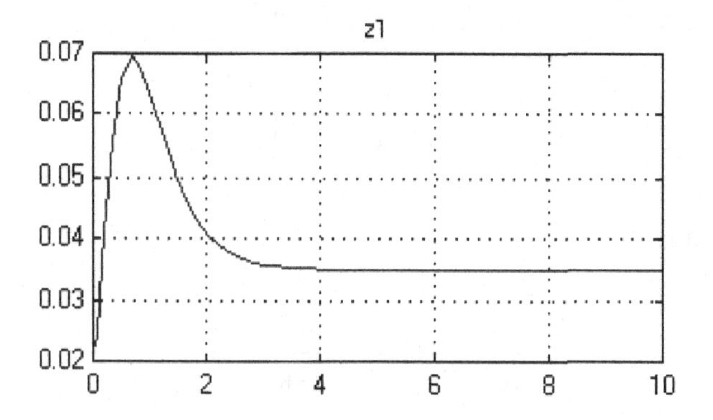

Fig. 8.6 Position $z_1 = x_1 = h(t)$ for the system control with feedback linearization

8.5 Nonlinear Controller of a Magnetic Suspension System

This controller includes state derivative feedback and is based on the input u and its time derivative [94] for a non-affine system with state vector x and scalar input

$$dx / dt = F(x, u)$$

Taking the time derivative of dx / dt = F(x, u)

$$\frac{d^2\mathbf{x}}{dt^2} = \frac{\partial F(\mathbf{x}, u)}{\partial x} \cdot \frac{\mathbf{dx}}{dt} + \frac{\partial \mathbf{F}(x, u)}{\delta u} \cdot \frac{du}{dt}$$

the control variable u obtained from the above equation as du/dt

$$\frac{du}{dt} = \left[\frac{\partial \mathbf{F}(x, u)}{\partial u}\right]^{-1} (v - \frac{\partial \mathbf{F}(\mathbf{x}, u)}{dx} \cdot \frac{dx}{dt}) \quad \text{for} \quad \frac{\partial \mathbf{F}}{du} \neq 0$$

where the new input v has the dimension of dx^2 / dt^2.

For the mechanical model of the nonlinear system given in state space format by

$$\dot{\mathbf{x}} = \begin{bmatrix} \dot{x}_1 \\ \dot{x}_2 \end{bmatrix} = \begin{bmatrix} x_2 \\ \dfrac{-k \cdot u^2}{m \cdot x_1^2} + g \end{bmatrix} = \mathbf{F}(\mathbf{x}, t)$$

the above equation for du/dt gives

$$\frac{du}{dt} = \frac{x_2}{x_1} \cdot u - \frac{m \cdot x_1^2}{2 \cdot k \cdot u} v$$

where v has the dimension of an acceleration.

The nonlinear controllers, given by this equation and the control variable u obtained above as a Nonlinear Input Transformation are restricted to conditions in which the variables u and x_1 in du/dt equation do not cross zero value. This is due to the fact that these variables appear in the denominator of the nonlinear control functions.

A controller consisting of PD feedback plus acceleration feedback, a PDA controller, can be chosen for obtaining the new control variable v

$$v = -K_a \cdot \left(\frac{d^2 x_1}{dt^2} - \frac{d^2 x_d}{dt^2}\right) - K_d \cdot \left(\frac{dx_1}{dt} - \frac{dx_d}{dt}\right) - K_p \cdot (x_1 - x_d)$$

where K_p, K_d and K_a are the position velocity and acceleration feedback error gains, respectively. Compared to the feedback linearization controller from Section 8.4, in this case a third state x_3 is not defined, but the resulting du/dt has to be integrated. The fact that no state transformation z = T(x) is required represents an important computation simplification.

Figure 8.8 contains the following blocks:

- PDA Control that generates v given states $x_1$and x_2 and state derivative dx_2/dt;
- Nonlinear Input Transformation with input v and the states $x_1$and x_2 and output du/dt;

- Nonlinear System with input u and output the states $x_1$and x_2 and state derivative dx_2/dt.

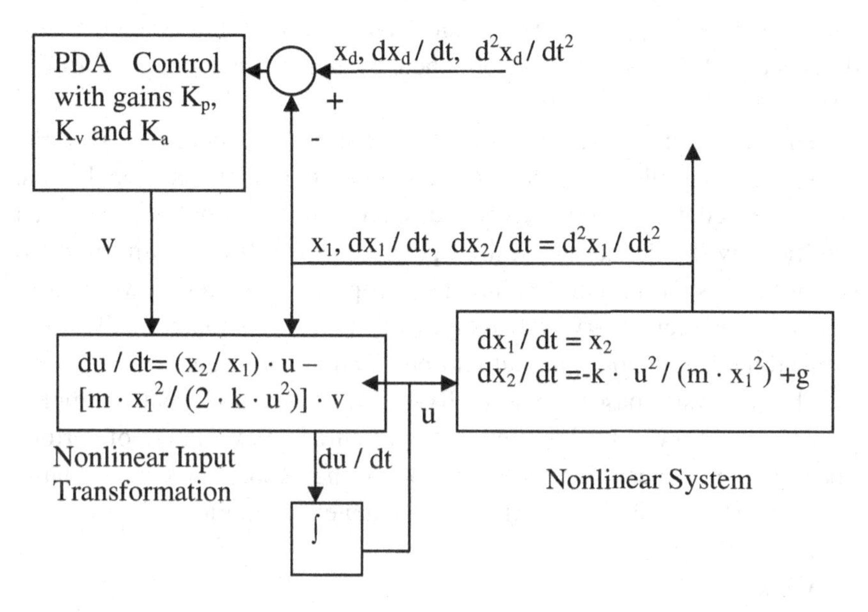

Fig. 8.8 Block diagram for PDA and nonlinear input v transformation controller

The SIMULINK simulation results for state derivative controller for step input of 0.035 [m] and initial condition h(0) = 0.0242 [m] are shown in Fig. 8.9.

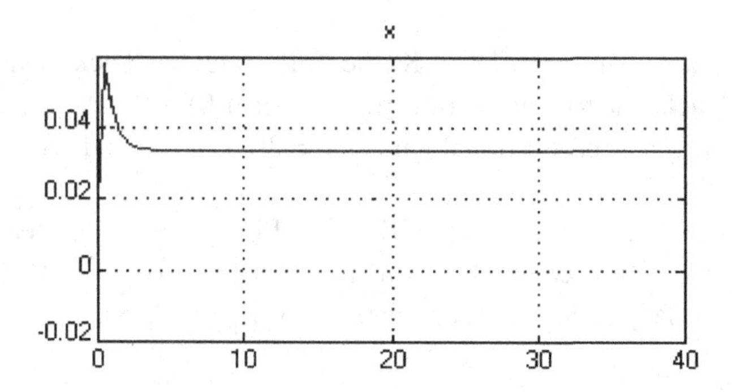

Fig. 8.9 Simulation results for state derivative controller

The results indicate that, again, the response corresponds to a stable system. The steady-state error is very small. The overshoot is high but the response shows, as in the case from section IV, a stable system, for a case in which a linear control would lead to an unstable system. Better transient performance can be obtained by a more elaborate controller design.

The results presented in this chapter permit to conclude that the nonlinear controllers, based on state-output linearization and state derivative control, while computationally more complex, increased significantly the domain of desired positions away from equilibrium for which the system remains stable. The proposed state derivative control has the supplementary advantage of being computationally less demanding than the on state-output linearization approach.

This analysis, based on a 1- DOF model, ignored many complex phenomena of multi-DOF systems but permitted an evaluation of various linear and nonlinear controllers to be further considered as valuable candidates for use in systems involving magnetic levitation.

Problems

1. Simulate using SIMULINK the PID control from Fig. 8.2 for proportional gain is chosen 125 and the PD control gains are chosen from root locus analysis with a zero at 25 and a pole at 60 and an integral control given by (1 + 1 / s), for a) h(0) = 0.04 [m] and x_d = 0.019 [m] and b) h(0) = 0.019 [m] and x_d = 0.04 [m].

2. Simulate using SIMULINK the full state feedback control and feedback linearization from Fig. 8.5 for h(0) = 0.019 [m] and x_d = 0.04 [m] and proportional gains $G_1 = 30$, $G_2 = 30$ and $G_3 = 0$.

3. Simulate using SIMULINK the PDA and nonlinear input transformation controller from Fig. 8.8 for h(0) = 0.019 [m] and x_d = 0.04 [m] and proportional gains $K_p = 30$, $K_v = 30$ and K_a= 0.0001.

Chapter 9

Inverse Problems Issues for Non-Minimum Phase Systems

9.1 Direct and Inverse Problems for Non-Minimum Phase Nonlinear Systems

9.1.1 *Introduction*

Feedback linearization is an inverse problem solution, used as a rigorous approach for the control of direct problems of non-linear systems. One of the most challenging difficulties in solving an inverse problem appears in the case that the direct problem corresponds to a non-minimum phase system. Even after successful design of a nonlinear controller, the issue of the non-minimum phase problem remains. A solution to solve the difficulties of inverse dynamics for such systems with non-minimum phase dynamics is output redefinition [75]. The output redefinition technique is formulated such that the resulting system to be inverted is a minimum phase system. This corresponds to a reduced order minimum phase approximation.

9.1.2 *Direct Problem for Non-Minimum Phase Systems*

If the non-minimum phase system is nonlinear, its linearization facilitates the calculation of the corresponding positive zeros in the direct problem formulation. The approach used here is Jacobian linearization.

Consider a non-minimum phase nonlinear time invariant system of the form

$$dx / dt = f(x) + g(x) \cdot u$$
$$y = h(x)$$

where state **x** **y**, **u** are the vectors of states, outputs and control, respectively. **f**(**x**), **g**(**x**) and **h**(**x**) are nonlinear functions and assumed to be smooth.

The Jacobian linearization of system about **x** = **0** is given by

$$dx / dt = A \cdot x + B \cdot u$$
$$y = C \cdot x$$

where

$$A = \delta f / \delta x |_{x=0}$$

$$B = g(0)$$

$$C\ A = \delta h / \delta x |_{x=0}$$

After a positive zero is found, matrix **C** is recalculated by moving positive zeros to the left half of s plane as **C*** using the output redefinition technique presented in [74]. The redefined output corresponds to a minimum phase system.

Assume the output rewritten is in form

$$y = h(x) = C \cdot x + h_{av}(x)$$

where $\mathbf{h}_{av}(\mathbf{x})$ is of order 2 or higher in **x**. The new output **y*** is

$$y^* = C^* \cdot x + h_{av}(x)$$

Consequently, the nonlinear system is approximated by a new system

$$dx / dt = f(x) + g(x) \cdot u$$
$$y^* = C^* \cdot x + h_{av}(x)$$

This system is a minimum phase system, and the feedback linearization control law can be applied to it to facilitate controller design.

9.1.3 *Neural Network Approach to Inverse Dynamics*

The complex calculations related to the output redefinition of system dynamics make a real-time computation very difficult. In order to facilitate the computation of the algorithm for real-time control, a neural networks approach was proposed [108, 112]. This method benefits from NARMA-L2 toolbox of MATLAB® which can be viewed as a neural network based feedback linearization tool. In NARMA-L2 controller, control inputs are computed algebraically, no on-line learning is needed at this stage any more and the computation required is greatly reduced compared with analytical feedback linearization [74, 107]. For this purpose, NARMA-L2 Neural Network is trained off-line to identify the forward dynamics with the redefined output, which is subsequently inverted to force the real output to approximately track a command input. Inverse dynamics can be obtained using neural networks. In a first stage, neural networks are trained to model the forward dynamics of an affine system. In the second stage, this neural network trained as direct problem is inverted to obtain an approximate inverse problem formulation. Simulation results for an Uninhibited Air Vehicle (UAV) illustrate the application of the proposed approach.

9.2 Feedback Linearization of a Non-Minimum Phase UAV

The performance of highly maneuverable UAV systems, requires enhanced operational capability in a constrained environment such as an air space containing static or moving obstacles. The high maneuverability of a UAV can be achieved by improving the flight control system using nonlinear control. Feedback linearization and dynamic inversion have been extensively applied in flight control especially in designing high maneuverable aircrafts and UAVs. [111]. Compared with traditional flight control design, which is often based on gain scheduling approach by dividing the flight space into linearizable subspaces, feedback linearization transforms the nonlinear dynamics of an aircraft into an equivalent linear system over the entire flight envelope, thus allowing us to use a global linear controller. However, the full-envelope nonlinear inversion of a UAV model is computationally intensive, because the UAV is a multi-input-multi-output (MIMO) system and it must be inverted in real time [113-115]. Furthermore, the exact input-output feedback linearization cannot be directly applied to a non-minimum phase UAV model.

To reduce the computational burden of the onboard computer, off-line trained neural network controllers were proposed to model the inverse dynamics of nonlinear systems [108, 112]. A direct way to do this is to train a neural network off-line to model the inverse dynamics of an aircraft using input-output pairs [113]. However, since the mathematic model of the inverse dynamic is not known a priori, the modeling errors could be significant. Therefore, adaptive control or on-line learning must be used to cancel out the modeling errors [114, 116]. An alternative way to apply inverse dynamics approach is to train neural networks to model the forward dynamics of the direct problem of an affine system, and then invert the neural network model to obtain an approximate inverse model, *i.e.* to obtain the formulation of the inverse problem of the system [117-119]. This method can use MATLAB® toolbox NARMA-L2, the neural network approach to feedback linearization. A set of neural networks that can be trained to approximate the Lie derivatives, such that the feedback linearization can be implemented step-by-step using these networks and, most importantly, the nonlinearity cancellation can be achieved [120]. Non-minimum phase systems result into an unstable system when subject directly to exact feedback inversion. A solution in this case is the so called approximate feedback linearization. This is done by approximating a non-minimum phase system with a minimum phase system, such that a bounded error tracking can be achieved [74]. In [120, 121], an approximate minimum phase model of a Vertical Takeoff and Landing (VTOL) aircraft was obtained by neglecting the coupling between the rolling moment and lateral acceleration. Similarly, for a slightly non-minimum phase Conventional Takeoff and Landing (CTOL) aircraft, ignoring some small forces, caused by control surfaces, from the equations of the system, will give a minimum phase model [122]. This method is only valid for slightly non-minimum phase systems and results in a loss of performance due to the un-modeled dynamics. Another method is output redefinition [74, 120]. In this case, the output is modified such that the resulting zero dynamics is stable. Output redefinition method has been successfully applied to control flexible manipulators. In [123], outputs are defined near the tip positions, such that the system becomes marginally minimum phase. In the field of flight control, the output redefinition method was also called Controlled (CV) Variable selection [124]. The selection of CVs is suitable for most conventional flight regimes. However, this selection may have to be modified for high-angle-of-attack or very-low-speed flight. Furthermore, the selection of CVs still relies to some extent on trial and error. In [125], the output is redefined using stable/anti-stable factorization performed on the zero dynamics of a discrete-time nonlinear non-minimum phase system. This is equivalent to moving the positive zero to the left half of s plane in continuous-time. This approach is however valid only for a class of non-minimum phase systems whose nonlinearities appear in output terms. In [126], a method is proposed to modify the output of the nonlinear aircraft model based on a transformation performed using the Jacobian linearization of the system. This transformation does not affect the left-half side zeros, thus the resulting system is essentially the same as the original one in the frequency range of interest. Using this approach, however, the system performance worsens when the frequency of desired output exceeds a certain limit. This limitation must be carefully considered in the context of designing tracking controllers for high maneuverable UAV. The non-minimum phase problem is still an active area of research in feedback control, given that all the methods mentioned above have their merits but also many limitations.

9.3 Mathematical Model for UAV Direct Problem

Nomenclature:

x_b, y_b, z_b	body axes
x_e, y_e, z_e	earth axes
ϕ, ψ, θ	bank angle, yaw angle, pitch angle
β	angle of sideslip
α	angle of attack

I_x, I_y, I_z	moments of inertia about body axes
p, q, r	angular velocity components along body axes
u, v, w	linear velocity components along b
δ_a, δ_r, δ_e	aileron, rudder, elevator deflections
W	weight of the aircraft
V	velocity of the aircraft
G	gravitational acceleration

It is important to establish which factor in the equations of a UAV model makes system behave non-minimum phase or contributes to positive zeros, and how the positive zeros (or the positive eigenvalues of the linearized zero-dynamics) change there values when the UAV is flying under different conditions. Positive zeros pose limitations on the frequency response of a UAV and hence affect its maneuverability.

The direct problem model has to be accurate enough to represent the needed features of the UAV dynamics, but also simple enough to be executed in real-time.

Denote:

$$\overline{l}_{\delta a} = l_{\delta a} + l_{\alpha\delta_a}\Delta\alpha$$

$$\overline{m}_{\alpha} = m_{\alpha} + m_{\dot{\alpha}}\Delta\alpha$$

$$\overline{m}_{q} = m_{q} + m_{\dot{\alpha}}$$

$$\overline{m}_{\delta e} = m_{\delta e} + m_{\dot{\alpha}} z_{\delta e}$$

$$\overline{n}_{\delta a} = n_{\delta a} + n_{\alpha\delta_a}\Delta\alpha$$

and the state vector $\boldsymbol{x}$ and the input $\boldsymbol{u}$ vector

$$\mathbf{x} = [p \quad q \quad r \quad \Delta\alpha \quad \beta \quad \phi \quad \theta]^T$$

$$\mathbf{u} = [\delta_a \quad \delta_r \quad \delta_e]^T$$

where δ_a, δ_r and δ_e denote the deflections of the aileron, the rudder and the elevator.

A direct nonlinear model of a small UAV is presented in detail in [108, 111, 112] and was used for simulations and controller design.

This direct problem model can be linearized into the canonic form [112]

$$d\mathbf{x} / dt = \mathbf{A} \cdot \mathbf{x} + \mathbf{B} \cdot \mathbf{u}$$
$$\mathbf{y} = \mathbf{C} \cdot \mathbf{x}$$

where

$$y_1 = \beta; \quad y_2 = \phi$$

The following [2 · 2] matrix of transform functions

$$\mathbf{H} = \mathbf{y}(s) / \mathbf{u}(s) = \begin{bmatrix} H_{11} & H_{12} \\ H_{21} & H_{22} \end{bmatrix}$$

can be obtained as

$$\mathbf{H}(s) = \mathbf{C} \cdot (s \cdot \mathbf{I} - \mathbf{A})^{-1} \cdot \mathbf{B}$$

where [108, 111, 112]

$$H_{11} = \frac{0.0071s^3 - 0.2488s^2 + 1.8533s - 0.3756}{s^4 - 4.364s^3 - 7.6707s^2 - 22.8411s - 0.0563}$$

$$H_{12} = \frac{6.3079s^2 + 25.2783s - 0.0902}{s^4 - 4.364s^3 - 7.6707s^2 - 22.8411s - 0.0563}$$

$$H_{21} = -\frac{45.83s^2 + 19.9397s + 271.1101}{s^4 - 4.364s^3 - 7.6707s^2 - 22.8411s - 0.0563}$$

$$H_{22} = -\frac{7.64s^2 + 4.1131s + 108.8293}{s^4 - 4.364s^3 - 7.6707s^2 - 22.8411s - 0.0563}$$

For an MIMO linear system, the zeros of H(s) can be calculated as the poles the inverse transfer function matrix $H^{-1}(s)$ of the system. Then a new matrix C^* can be determined such that the resulting system has no positive zeros. To find the zeros of the system, one may invert the transfer matrix H and consider its poles. The inverse transfer matrix H^{-1} is given by

$$\mathbf{H}^{-1} = \begin{bmatrix} H_{11}^{(inv)} & H_{12}^{(inv)} \\ H_{21}^{(inv)} & H_{22}^{(inv)} \end{bmatrix}$$

where,

$$H_{11}^{(inv)} = \frac{7.64s^2 + 4.1131s + 108.8293}{0.0542(s - 5368.31)}$$

$$H_{12}^{(inv)} = \frac{6.3079s^2 + 25.2783s - 0.0902}{0.0542(s - 5368.31)}$$

$$H_{21}^{(inv)} = -\frac{45.83s^2 + 19.9397s + 271.1101}{0.0542(s - 5368.31)}$$

$$H_{22}^{(inv)} = -\frac{0.0071s^3 - 0.2488s^2 + 1.8533s - 0.3756}{0.0542(s - 5368.31)}$$

Therefore, the zero of the direct problem model **H**(s) is at s = 5368.31 = b which coincides with the positive eigenvalue of the zero-dynamics of the UAV nonlinear model. The new system output **y*** can now be calculated using output redefinition to eliminate positive zeros from the direct problem. For a system with outputs $\mathbf{y}_i(s)$, i = 1, 2,..., m, first a constant matrix **M** is defined as link between actual output **y** and the redefined output y* using the equation for **y**(s)

$$\begin{bmatrix} y_1(s) \\ y_2(s) \\ \vdots \\ y_m(s) \end{bmatrix} = M \begin{bmatrix} y_1^*(s) - (s/b)y_1^*(s) \\ y_2^*(s) \\ \vdots \\ y_m^*(s) \end{bmatrix}$$

where **M** has to satisfy the constraint

$$\mathbf{M}^{-1} \cdot \mathbf{H}(b) = \begin{bmatrix} 0 & 0 & \cdots & 0 \\ * & * & \cdots & * \\ \vdots & \vdots & \ddots & \vdots \\ * & * & \cdots & * \end{bmatrix}$$

Since the value of the transfer matrix **H**(s) at s = 5368.31 = b is

$$\mathbf{H}(b) = \begin{bmatrix} 1.3129 & 0.2189 \\ -1.5891 & -0.2649 \end{bmatrix}$$

the constant matrix $\mathbf{M}^{-1}$ can be obtained as

$$\mathbf{M}^{-1} = \begin{bmatrix} 1 & 0.8262 \\ 0 & 1 \end{bmatrix}$$

A redefined transfer functions matrix can now be obtained without any positive zeros. This is the result of redefining the output such that the positive zero from **H**(s)

a) does not appear in the new matrix **H***(s)

b) is moved to the left hand side of the s-plane for **h***(s)

a) The following equation shows the relationship between **H**(s) and the new **H***(s) with the positive zero removed

$$\mathbf{H}(s) = \mathbf{M}\begin{bmatrix} 1-\dfrac{s}{b} & 0 & \cdots & 0 \\ 0 & 1 & \cdots & 0 \\ \vdots & \vdots & \ddots & \vdots \\ 0 & 0 & \cdots & 1 \end{bmatrix} \cdot \mathbf{H}^{*}(s)$$

where

$$H_{11}^{*} = -\frac{38.115s^2 + 14.6621s + 224.3589}{s^4 - 4.364s^3 - 7.6707s^2 - 22.8411s - 0.0563}$$

$$H_{12}^{*} = \frac{21.8634s - 90.0019}{s^4 - 4.364s^3 - 7.6707s^2 - 22.8411s - 0.0563}$$

$$H_{21}^{*} = -\frac{45.83s^2 + 19.9397s + 271.1101}{s^4 + 4.364s^3 + 7.6707s^2 + 22.8411s + 0.0563}$$

$$H_{22}^{*} = -\frac{7.64s^2 + 4.1131s + 108.8293}{s^4 + 4.364s^3 + 7.6707s^2 + 22.8411s + 0.0563}$$

The new output matrix **C*** is computed using

$$\mathbf{C}^* = \mathbf{H}^*(s) \cdot [(s\mathbf{I} - \mathbf{A})^{-1} \cdot \mathbf{B}]^{-1}$$

that gives

$$\mathbf{C}^* = \begin{bmatrix} 1.5866\times10^{-4} & 0 & -1.862\times10^{-4} & 0 & 1 & 0.82618 & 2.3455\times10^{-4} \\ 0 & 0 & 0 & 0 & 0 & 1 & 0 \end{bmatrix}$$

This leads to the new output after removing the positive zero b from **H**(s)

b) Using equation,

$$\mathbf{h}^*(s) = \begin{bmatrix} \frac{1+s/b}{1-s/b} & 0 & \cdots & 0 \\ 0 & 1 & \cdots & 0 \\ \vdots & \vdots & \ddots & \vdots \\ 0 & 0 & \cdots & 1 \end{bmatrix} \cdot \mathbf{M}^{-1} \cdot \mathbf{H}(s)$$

the transfer matrix $\mathbf{h}^*(s)$ results from moving the positive zero to the left hand side of the s-plane. For b = 5368.31

$$\mathbf{h}^*(s) = \begin{bmatrix} h_{11}^* & h_{12}^* \\ h_{21}^* & h_{22}^* \end{bmatrix} = \begin{bmatrix} \frac{1+(s/5368.31)}{1-(s/5368.31)} & 0 & \cdots & 0 \\ 0 & 1 & \cdots & 0 \\ \vdots & \vdots & \ddots & \vdots \\ 0 & 0 & \cdots & 1 \end{bmatrix} \cdot \mathbf{M}^{-1} \cdot \mathbf{H}(s)$$

where,

$$h_{11}^* = -\frac{0.0071s^3 + 38.1177s^2 + 14.7039s + 224.3589}{s^4 + 4.364s^3 + 7.6707s^2 + 22.8411s + 0.0563}$$

$$h_{12}^{*} = \frac{0.0041s^2 + 21.8468s - 90.0023}{s^4 + 4.364s^3 + 7.6707s^2 + 22.8411s + 0.0563}$$

$$h_{21}^{*} = -\frac{45.83s^2 + 19.9397s + 271.1101}{s^4 - 4.364s^3 - 7.6707s^2 - 22.8411s - 0.0563}$$

$$h_{22}^{*} = -\frac{7.64s^2 + 4.1131s + 108.8293}{s^4 + 4.364s^3 + 7.6707s^2 + 22.8411s + 0.0563}$$

The new output matrix **c*** could be obtained as

$$\mathbf{c}^{*} = \begin{bmatrix} 3.2053 \cdot 10^{-4} & 0 & -3.7617 \cdot 10^{-4} & 0 & 1 & 0.82619 & 0 \\ 0 & 0 & 0 & 0 & 0 & 1 & 0 \end{bmatrix}$$

A new transfer function results

$$\frac{y*(s)}{\delta_a(s)} = \frac{-0.3756}{s^4 + 4.364s^3 + 7.6707s^2 + 22.8411s + 0.0563}$$

The output for UAV model is redefined by moving the positive zeros to the left half of s plane, so that these two systems with new outputs are now minimum phase and can be inverted.

9.4 Simulation Results for the Neural Controller and Output Redefinition

In this section, the NARMA-L2 neural controller and the output redefinition technique are applied to a nonlinear non-minimum phase UAV system [108, 112, 130].

The control scheme for the UAV, shown in Fig. 9.1 is based on the inverse problem solution, i.e. the nonlinear inversion of a UAV model in real time.

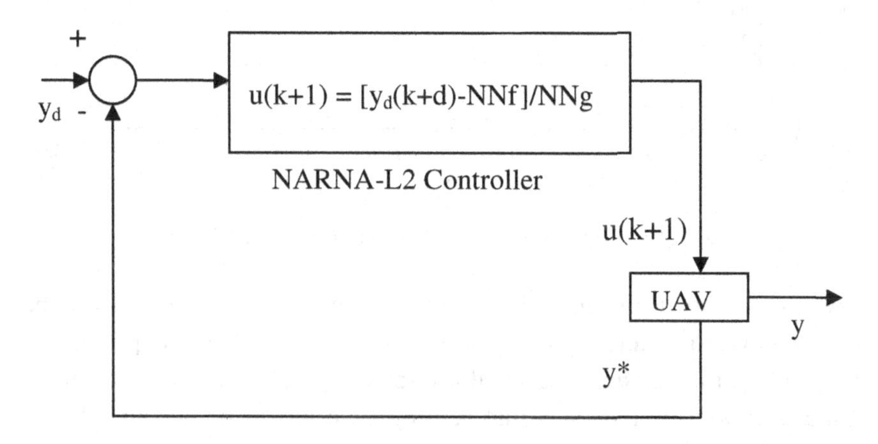

Fig. 9.1 UAV Control System based on NARMA-L2

To reduce the significant computation burden of the onboard computer of a UAV, an approximate feedback linearization methodology using offline trained NARMA-L2 neural networks NNf and NNg is analyzed here. The NARMA-L2 neural network is trained off-line for the direct problem, to model the forward dynamics of a UAV, with a redefined output of the UAV model. Then, the trained neural network (NN) is inverted so that it cancels out the nonlinearities of the UAV model. NORMA-L2 Controller has as inputs the command inputs y_d and the UAV model output at time step k and generates the control command for the next time step k+1 using the output of the direct NNf and the inverse NNg. This becomes the Control Input to the UAV model which forces the output y to approximately track the desired output y_d.

The results are shown as dotted lines for the command input and continuous line for the actual output. Figure 9.2 shows the response to a step input with a settling time of approx. 20 [s].

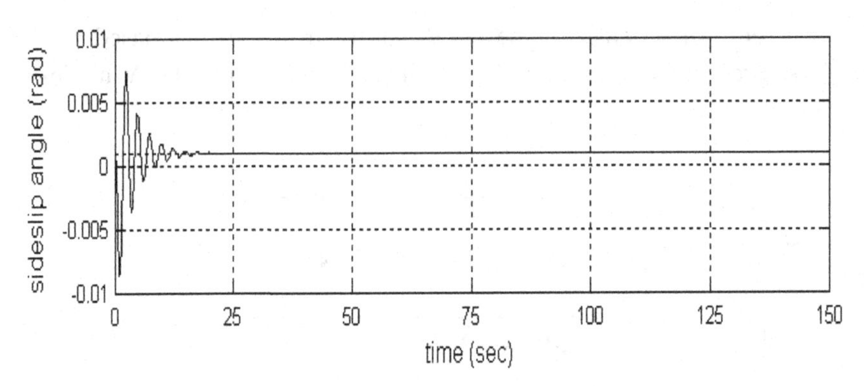

Fig. 9.2 The sideslip angle response to a step command input

The results shown in Fig. 9.3, for square wave input, show that for the square wave command input period which decreases from 160 [s] in a) to 20 [s] in c), the transient regime does not vanish before a new command value occurs. This indicates that settling time has to be accounted as a limit to the period of the changes in the command input.

These results exemplify a solution to the non-minimum phase problem in UAV control design. Since the feedback linearization cannot be directly applied to a UAV, which is non-minimum phase, the output redefinition approach redefines the output of the system, such that the resulting new system has stable zero-dynamics and the inverse of the new system is asymptotically stable.

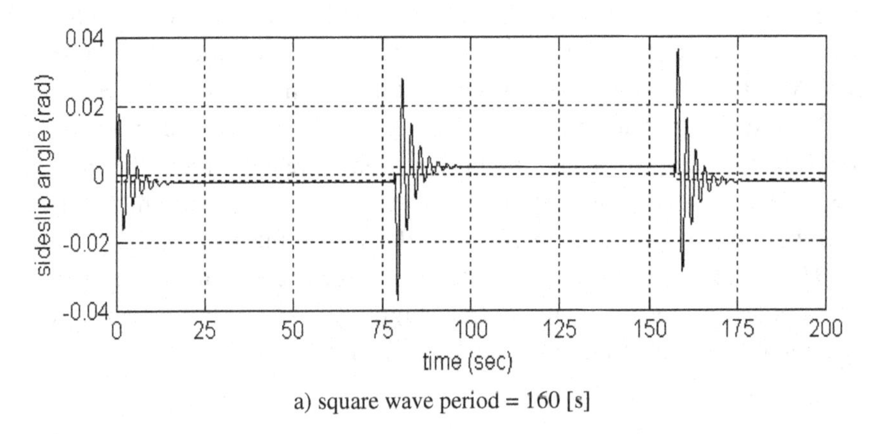

a) square wave period = 160 [s]

Fig. 9.3 Sideslip angle responses to square wave command inputs

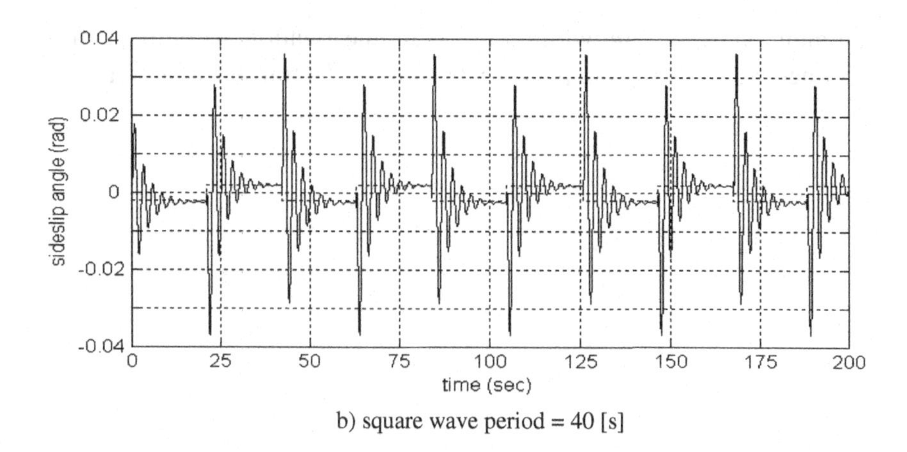

b) square wave period = 40 [s]

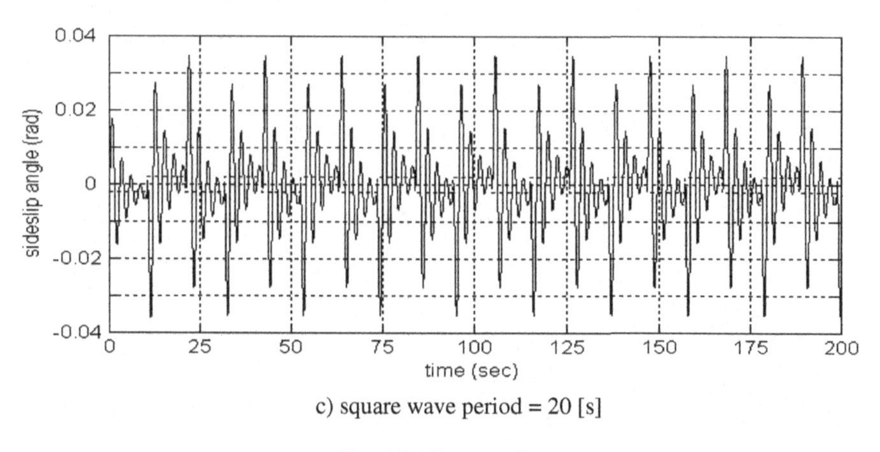

c) square wave period = 20 [s]

Fig. 9.3 (*Continued*)

In nonlinear case, due to the impossibility to define the transfer function, output redefinition cannot be directly implemented. To solve this problem, one may first calculate the Jacobian linearization of the nonlinear system, and then apply above techniques to the linearized system to define the new outputs. As a result, for the nonlinear non-minimum phase system, with its Jacobian linearization, its output can be redefined by removing the positive zeros from the transfer function.

References

[1] H. Bühler, Mecatronique, EPFL, 1992.

[2] W. Stadler, Analytical Robotics and Mechatronics, McGraw-Hill, 1995.

[3] D.M. Auslander, C.J. Kempf, Mechatronics: Mechanical System Interfacing, Prentice Hall, 1996.

[4] D. Shetty, R. Kolk, Mechatronics System Design, PWS, 1997.

[5] W. Bolton, Mechatronics, Addison Wesley Longman, 1999.

[6] D. Popovic and L. Vlacic, Mechatronics in Engineering Design and Product Development, M. Dekker, 1999.

[7] M. Histand and D. Alciatore, Mechatronics and Measurement Systems, McGraw-Hill, 1999.

[8] D. Necsulescu, Mechatronics, Prentice Hall, 2002.

[9] R. Bishop (Editor-in-Chief), The Mechatronics Handbook, CRC Press, 2002.

[10] C. Doumanidis, Thermomechatronics: Mechatronic Systems and Guidance Techniques in Thermal Manufacturing Process Control. Ch. 5 in Mechatronic Systems Techniques and Applications, Vol. 1, Industrial Manufacturing, (editor C. Leondes), Gordon and Breach Science Publishers, 2000.

[11] D.A. Wells, Lagrangian Dynamics, Schaum's Outline Series, McGraw-Hill, 1967.

[12] D. Halliday and R. Resnick, Fundamentals of Physics, J. Wiley, 1981.

[13] F. Bueche and E. Hecht, College Physics, Schaum's Outline Series, McGraw-Hill, 1997.

[14] E. Kreyszig, Advanced Engineering Mathematics, J. Wiley, 1988.

[15] M. Spiegel, Applied Differential Equations, Prentice Hall, 1967.

[16] P. O'Neil, Advanced Calculus, Macmillan, 1975.

[17] D. Inman, Engineering Vibration, Prentice Hall, 1994.

[18] W. Harmon Ray, Advanced Process Control, McGraw-Hill, 1981.

[19] P. Albertos and A. Sada, Multivariable Control Systems, Springer, 2004.

[20] R. Baican, D. Necsulescu, Applied Virtual Instrumentation, WIT Press, 2000.

[21] L.E. Kinsler et al., Fundamentals of Acoustics, J. Wiley, 2000.

[22] D.A. Murio, The Mollification Method and the Numerical Solution of Ill-Posed Problems, J. Wiley, 1993.

[23] C. Hansen and S. Snyder, Active Control of Noise and Vibration, Chapman & Hall, 1997.

[24] M. Spiegel, Advanced Mathematics for Engineers and Scientists, Schaum's Outline Series, McGraw-Hill, 1971.

[25] D. Betounes, Partial Differential Equations for Computational Science with Maple and Vector Analysis, Springer, 1998.

[26] J. Tuma, Handbook of Numerical Calculations in Engineering, McGraw-Hill, 1989.

[27] D. Necsulescu, Object oriented modeling of a smart structure, in Computational Methods for Smart Structures and Materials, Editors C. Brebbia and A. Samartin, WIT Press, 2000, pp. 53-62.

[28] C. Gerald, Applied Numerical Analysis, Addison-Wesley, 1980.

[29] R. Green, Classical Mechanics with Maple, Springer, 1995.

[30] J. Beck et al., Inverse Heat Conduction. Ill problems, Wiley, 1985.

[31] L. Kaplan, Emergency and Disaster Planning Manual, McGraw-Hill, 1996.

[32] Proc. of the Workshop on Critical Infrastructure Protection and Emergency Preparedness, Ottawa, 10-11 Dec. 2004.

[33] G. Jiang et al., Infrastructure Web: Distributed Monitoring and Managing Critical Infrastructures, Proc. of SPIE, V. 4232, 2001, pp. 104-114.

[34] T.C. Hutchinson and F. Kuester, Monitoring Global Earthquake-Induced Demands using Vision-Based Sensors, IEEE Trans Instrumentation and Measurement, Vol. 53, Nr. 1, Febr. 2004, pp. 80-84.

[35] B. Ando, M. Coltelli and M. Sambataro, A Measurement Tool for Investigating Cooling Lava Properties, IEEE Trans Instrumentation and Measurement, Vol. 53, Nr. 2, Apr. 2004, pp. 507-513.

[36] D. Necsulescu and G. Ganapathy, Inverse Problems in Infrastructure Sensing for Emergency Preparedness, Proc. of the Workshop on Cyber Infrastructure – Emergency Preparedness Aspects, Ottawa, 21-22 Apr 2005, pp. 71-75.

[37] A. Schiff, Seismic Design for Selected Equipment in New or Refurbished Substations, Proc. Disaster Preparedness Conf III, 1994, pp. 4-2 to 4-10.

[38] D. Wilson, et al., Chemical Sensors for Portable, Handheld Instruments, IEEE Sensors Journal, Vol. 1, No. 4, 2001, pp. 256-274.

[39] A. Zoubir et al., Signal Processing Techniques for Landmine Detection using Ground Penetrating Radar, IEEE Sensors Journal, Vol. 2, No. 1, 2002, pp. 41-51.

[40] D. Broomhead and R. Jones, Condition Monitoring and Failure Prediction in Chaos, IEEE Conf. on Adv. Vibration Measurements, 1992.

[41] A. Kirsch, An Introduction to the Mathematical Theory of Inverse Problems, Springer, 1996.

[42] D. Campos-Delgado et al., Thermoacoustic Instabilities Modeling and Control, IEEE Trans. Control Systems Technology, Vol. 11, No. 4, July 2003, pp. 429-447.

[43] P. Charboneaud, C. Carrillo and S. Medar, Robust Control Reconfiguration of a Thermal Process with Multiple Operating Modes, IEEE Trans. Control Systems Technology, Vol. 11, No. 4, July 2003, pp. 529-447.

[44] C. Doumanidis, Thermomechatronics, in C. Leondes (Ed), Mechatronic Systems Techniques and Applications, Vol. 1, Industrial Manufacturing, Gordon and Breach Science Publishers, 2000, pp. 255-313.

[45] S. Twomey, Introduction to the Mathematics of Inversion in Remote Sensing and Indirect Measurements, Elsevier Scientific Publishing Company, 1977.

[46] J. Beck, Sequential Methods in Parameter Estimation, Ch. 1 in Inverse Engineering Handbook (edited by K. A. Woodbury), CRC Press, 2003.

[47] D. Murio, Mollification and Space Marching, Ch. 4 in Inverse Engineering Handbook (edited by K. A. Woodbury), CRC Press, 2003.

[48] Y. Jarny, The Adjoint Method to Compute the Numerical Solutions of Inverse Problems, Ch. 4 in Inverse Engineering Handbook (edited by K. A. Woodbury), CRC Press, 2003.

[49] W. Press et al., Numerical recipes in Fortran, Cambridge University Press, 1986.

[50] K. Ogata, Modern Control Engineering, 4th edition, Prentice Hall, 2000.

[51] D.S. Necsulescu, R.F. De Abreu, F. Bakhtiari-Nejad, Smart Structure Integrity Monitoring using Transient Response, Computational Methods for Smart Structures and Materials, P. Santini, M. Marchetti, C.A. Brebbia (Ed), WIT Press, 1998, pp. 141-150.

[52] D. Necsulescu, R. De Abreu, Dynamic Model Based Monitoring of Space Structures Integrity, Proc. 4th Int. Conf. Dynamics and Control of Structures in Space, Cranfield, UK, 24-28 May, 1999.

[53] D. Necsulescu, F.Bakhtiari-Nejad, Kuoc-Vai Iong, Transient Vibration Response Method for Non-destructive Testing of Smart Structures, Technical report, Department of Mechanical Engineering, University of Ottawa, 1998.

[54] K. Ogata, Model Control Engineering, Prentice Hall, 2002.

[55] R. Szilard, Theory and Analysis of Plates, Prentice Hall, 1974.
[56] FEMLAB 3 Modeling Guide, 2004.
[57] FEMLAB 3 Model Library, 2004.
[58] Y. Zheng, R. Gourban and M. El-Tanany, Experimental Evaluation of a Nested Microphone Array with Adaptive Noise Cancellation, IEEE Trans. Instr. and Measurement, Vol. 53, No. 3, June 2004, pp. 777-786.
[59] J. Foreman, Sound Analysis and Noise Control, Van Nostrand, 1990.
[60] L. Beranek and I. Ver, (eds) Noise and Vibration Control Engineering, Wiley, 1992.
[61] S.S. Blackman, Association and Fusion of Multiple Sensor Data, in Multitarget-Multisensor Tracking, (Y. Bar-Shalom Editor), Artech House, 1990.
[62] J.A. Richards and X. Jia, Remote Sensing Digital Image Analysis: An Introduction, Springer, 2006.
[63] A.N. Tikhonov and V.Y. Arsenin, Solutions to Ill-posed Problems, John Wiley, 1977.
[64] M. Bortero, C. De Mol and E.R. Pike, Applied Inverse Problems in Optics, in Inverse and Ill-posed Problems (edited by H. Engl and C. Groetsch), Academic Press, 1987.
[65] W. Menke, Geophysical Data Analysis: Discrete Inverse Theory, Academic Press, 1989.
[66] A.G. Ramm, An Inverse Problem in Ocean Acoustics, Journal of Inverse Ill-Posed Problems, Vol. 9, No. 1, 2001, pp. 95-102.
[67] J.C. Santamarina and D. Fratta, Introduction to Discrete Signals and Inverse Problems in Civil Engineering, ASCE Press, 1998.
[68] P.M. Mather, Computer Processing of Remotely-Sensed Images, John Wiley, 2004
[69] S.M. de Jong and F.D. van der Meer, (Editors), Remote Sensing Image Analysis, (Springer, 2006).
[70] G. Franklin, J. Powell, A. Emami-Naeini, Feedback Control of Dynamic Systems, Prentice-Hall, 2006.
[71] E. Doebelin, Measurement Systems, McGraw-Hill, 1990.
[72] B. Friedlander, Control System Design, McGraw-Hill, 1986.
[73] U.M. Sultangazin, Mathematical Problems Connected with Construction of Algorithm for Atmosphere Correction in Remote Sensing, Journal Inverse Ill-Posed Problems, Vol. 9, No. 6, 2001, pp. 655-668.
[74] J-J. Slotine and W. Li, Applied Nonlinear Control, Prentice Hall, 1991.

[75] L. Jackson, Digital Filters and Signal Processing, Kluwer Academic Publishers, 1986.

[76] A. Tarantola, Inverse Problem Theory, SIAM, 2005.

[77] V. Topa, Optimal Design of the Electromagnetic Devices using Numerical Methos, VUB University Press, 2000.

[78] K. Woodbury, Sequential Function Specification Method, Ch. 2 in Inverse Engineering Handbook (edited by K. A. Woodbury), CRC Press, 2003.

[79] I. Zeldovitch and A. Myschkis, Elements de mathematiques appliqués, Editions MIR, 1974.

[80] J.C. Willems, The Behavioral Approach to Open and Interconnected Systems, IEEE Control Systems Magazine, De. 2007, pp. 46-99.

[81] E. Part-Enander et al., The MATLAB® Handbook, Addison-Wesley, 1996.

[82] H. Kuttruff, *Room Acoustics*, Third Edition, Elsevier Applied Science, 1991.

[83] Malcolm J. Crocker, *Handbook of Acoustics*, Wiley Interscience, 1998.

[84] J.B. Allen and David A. Berkley, Image method for efficiently simulating small-room acoustics, *J. Acoust. Soc. Am.* 65, 943-950, 1979.

[85] www.acoustics.salford.ac.uk/student_area/bsc3/room_acoustics/Geo-2003.pdf

[86] B.M. Gibbs, D.K. Jones, A Simple Image Method for Calculating the Distribution of Sound Pressure Level within an Enclosure, *Acustica* 26, 24-32, 1972.

[87] http://freesound.iua.upf.edu/index.php

[88] D. Necsulescu, W. Zhang, W. Weiss and J. Sasiadek, Room Acoustics Measurement System Design using Simulation And Experimental Studies, IEEE Transactions Instrumentation and Measurement, (accepted for publication on 19 Apr., 2008).

[89] D. Necsulescu, W. Zhang and W. Weiss, The Design of Measurement System for Room Acoustics, Buletin Inst Politehnic Iasi, Vol. LII, Fasc. 6B, 2006, pp. 53-58.

[90] M. Stieber, E. Petriu and G. Vukovich, Systematic Design of Instrumentation for Mechanical Systems, IEEE Trans Instrumentation and Measurement, Vol. 45, Nr. 2, Apr 1996, pp. 406-412.

[91] A. Schiff, Seismic Design for Selected Equipment in New or Refurbished Substations, Proc. Disaster Preparedness Conf III, 1994, pp. 4-2 to 4-10.

[92] F. Mondinelli and Z. Kovacs-Vajna, Self-Localizing Sensor Network Architectures, IEEE Trans Instrumentation and Measurement, Vol. 45, Nr. 2, Apr 1996, pp. 277-283.

[93] D.D. Ang et al., Moment Theory and Some Inverse Problems in Potential Theory and Heat Conduction, Springer, 2002.

[94] D.S. Necsulescu, M. Ceru, Nonlinear Control of Magnetic Bearings, Journal of Electrical Engineering, Vol. 2, Nr. 1, 2002, Art no. 14.

[95] R. Williams, Digital Control of Active Bearings, IEEE Trans on Ind. Electronics, Feb. 1990, pp. 19-27.

[96] J.H. Tarn and J.Y. Juang, Time-Delay Control of Magnetic Levitated Linear Positioning System, Proc ACC, June 1994, pp. 1947-1951.

[97] H-S. Jeong and C-W. Lee, Time Delay Control with State Feedback for Azimuth Motion of the Frictionless Positioning Device, IEEE Trans Mechatronics, Sept 1997, pp. 161-168.

[98] K. Youcef-Toumi and S-T. Wu, Input/Output Linearization using Time Delay Control, ASME Journal of Dynamic Systems, Measurement and Control, March 1002.

[99] Y. Zhuravlyov, On LQ Control of Magnetic Bearings, IEEE Trans Control Systems Technol, March 2000, pp. 344-350.

[100] S. Ueno, Characteristics and Control of a Bi-directional Axial Gap Combined MotorBearing, IEEE Trans Mechatronics, Sept 2000, pp. 310-318.

[101] S. Mittal and C-H. Menq, Hysteresis Compensation in Electromagnetic Actuators through Preisach Model Inversion, IEEE Trans on Mechatronics, Dec 2000, pp. 394-409.

[102] Z. Ji et al., Contactless Magnetic Leadscrew: Modeling and Load Determination, Proc. ACC, June 2000, pp. 1062-1066.

[103] Y.C. Kim and K.H. Kim, Gain Scheduled Control of Magnetic Suspension System, Proc. ACC, June 1994, pp. 3127-3131.

[104] Revolve Technologies, Magnetic Bearings Systems, CSME Bulletin, May 2000, pp. 5-11.

[105] E.B. Magrab et al., An Engineer's Guide to MATLAB, Prentice Hall, 1999.

[106] P.K. Sinha, Electromagnetic Suspension: Dynamics and Control, Peter Peregrinus, 1987.

[107] MATLAB™ and SIMULINK™, The MATHWORKS, 1998.

[108] Y. Jiang, D. Necsulescu and J. Sasiadek, "Robotic Unnmanned Aerial Vehicle Trajectory Tracking Control", SYROCO 2006 September 6-8, 2006, Bologna.

[109] H.T. Banks, Inverse problems for distributed parameter systems, Proceedings of the 29th IEEE Conference on Decision and Control, 1990, Vol. 1, Issue, 5-7 Dec 1990, pp. 13-16.

[110] K. van den Doel et al., Dynamic level set regularization for large distributed parameter estimation problems, Inverse Problems, 23, 2007, pp. 1271-1288.

[111] D. Enns, Bugajski, D.J., Hendrick, R.C., and Stein, G. "Dynamic Inversion: An Evolving Methodology for Flight Control Design", Int. J. Control, 1994, Vol. 59, No. 1, 71-91.

[112] Y. Jiang, Neural Network Based Linearization Control of an Unmanned Aerial Vehicle, MASc Thesis, University of Ottawa, Nov. 2005.

[113] B.S. Kim and Calise, A J. Nonlinear Flight Control Using Neural Networks, Journal of Guidance, Control and Dynamics, Vol. 20, No. 1, January-February 1997.

[114] L. Yan and Li, C.J. Robot learning control based on recurrent neural network inverse model, J. Robot. Syst., Vol. 14, No. 3, pp. 199-212, 1997.

[115] G.L. Plett, Adaptive Inverse Control of Linear and Nonlinear Systems Using Dynamic Neural Networks, IEEE Transactions on Neural Networks, Vol. 14, No. 2, March 2003.

[116] A. Green and Sasiadek, J., Adaptive Control of a Flexible Robot Using Fuzzy Logic, AIAA Journal of Guidance, Control, and Dynamics, Vol. 28, No. 1, January-February 2005.

[117] Z. Su and Khorasani, K. "A Neural Network Based Controller for a Single-Link Flexible Manipulator Using the Inverse Dynamics Approach", IEEE Transactions on Industrial Electronics, Vol. 48, No. 6, December 2001.

[118] K.S. Narendra and Mukhopadhyay, S. "Adaptive Control Using Neural Networks and Approximate Models", IEEE Transactions on Neural Networks, Vol. 8, No. 3, May 1997.

[119] S. He, Reif, K. and Unbehauen, R. "A Neural Approach for Control of Nonlinear Systems with Feedback Linearization", IEEE Transactions on Neural Networks, Vol. 9, No. 6, November 1998.

[120] J. Hauser, Sastry, S., Meyer, G. "Nonlinear Control Design for Slightly Non-minimum Phase Systems: Application to V/STOL Aircraft", Automatica, Vol. 28, No. 4, pp. 665-679, 1992.

[121] M. Oishi and Tomlin, C. "Switching in Non-minimum Phase Systems: Applications to a VSTOL Aircraft", in Proceedings of the American Control Conference, 2000.

[122] J.J. Romano and Singh, S.N. "I-O Map Inversion, Zero Dynamics and Flight Control", IEEE Transactions on Aerospace and Electronic Systems, Vol. 26, No. 6, November 1990.

[123] M. Moallem, Khorasani, K. and Patel, R.V. "An Inverse Dynamics Slidin Control Technique for Flexible Multi-Link Manipulators", Proceedings of the American Control Conference, Albuquerque, New Mexico, June 1997.

[124] B. L. Stevens and Lewis, F.L. Aircraft Control and Simulation, Second Edition, Wiley, 2003.

[125] H.A. Talebi, Khorasani, K. and Patel, R.V. "Experimental Results on Tracking Control of a Flexible-Link Manipulator: A New Output Redefinition Approach", Proceedings of the 1999 IEEE International Conference on Robotics and Automation, Detroit, Michigan, May 1999.

[126] L. Benvenuti and Benedetto, M.D.DI. "Approximate Output Tracking for Nonlinear Non-minimum Phase Systems with An Application to Flight Control", International Journal of Robust and Nonlinear Control, Vol. 4, 397-414, 1994.

[127] B. Friedland, Advanced Control System Design, Prentice Hall, 1996.

[128] M.R. Golbahar Haghighi, D.S. Necsulescu, M. Eghtesad, P. Malekzadehc, Two-dimensional Inverse Analysis of Functionally Graded Materials in Estimating Time-Dependent Surface Heat Flux, Numerical Heat Transfer, Part A. (accepted for publication on 25 June, 2008).

[129] K. Wen, D. Necsulescu, J. Sasiadek, Haptic Force Control based on Impedance / Admitance Control Aided by Visual Feedback, International Journal of Multimedia Applications, 37, Number 1, March 2008, pp. 39-52.

[130] D. Necsulescu, Yi-Wu Jiang and B. Kim, Neural Network Based Feedback Linearization Control of an Unmanned Aerial Vehicle, International Journal of Automation and Computing, Vol. 1, Jan 2007, pp. 71-79.

Index

About the Author

Dr. Dan S. NECSULESCU is professor in the Department of Mechanical Engineering and Director of the Engineering Management Program at the University of Ottawa.

He obtained his diploma and PhD in Engineering from Polytechnic University of Bucharest, Bucharest and Licentiate in Philosophy from the University of Bucharest, Bucharest, Romania

He worked as design and research engineer and nuclear safety engineer until 1980, when he joined the University of Ottawa. He chaired the Mechanical Engineering Department from 1990 to 1993. Since 1996-present, he is director of the Engineering Management Program.

He has published two books — *Applied Virtual Instrumentation* with R. Baican (WIT Press) in 2000 and *Mechatronics* (Prentice Hall) in 2002.

His research interests include Control and Dynamics, Mechatronics, Robotics, Formations of Autonomous Vehicles and Self-Organizing Engineered Systems.